William Mellen Cate

The Child of Promise

Or, the Isaac of medicine and Ishmael, the half brother. Being a comprehensive

glance at the instincts and predilections of the rival schools of medicine

William Mellen Cate

The Child of Promise
Or, the Isaac of medicine and Ishmael, the half brother. Being a comprehensive glance at the instincts and predilections of the rival schools of medicine

ISBN/EAN: 9783337290788

Printed in Europe, USA, Canada, Australia, Japan

Cover: Foto ©berggeist007 / pixelio.de

More available books at **www.hansebooks.com**

THE CHILD OF PROMISE;

OR,

ISAAC OF MEDICINE

AND

ISHMAEL, THE HALF BROTHER,

BEING A

COMPREHENSIVE GLANCE AT THE INSTINCTS AND
PREDILECTIONS OF THE

RIVAL SCHOOLS OF MEDICINE.

BY

WILLIAM MELLEN CAT

*Doctor of Medicine ; formerly Professor of Chemistry and Toxicology in the
Homœopathic Medical College of Missouri ; Member of the American
Institute of Homœopathy ; of the Massachusetts Homœopathic
Medical Society ; of the Massachusetts Surgical and
Gynæcological Society, etc., etc.*

WASHINGTON, D. C.
H. B. BURNHAM & CO.

And he shall dwell with wild men. " His hand will be against every man, and every man's hand against him."

'For Ishmael" "I have blessed him, and will make :ful, and will multiply him exceedingly ;" " But enant will I establish with Isaac," " for an ever-lasting covenant and with his seed after him."

"And Sarah saw the son of Hagar which she had borne unto Abraham MOCKING ! !

" Wherefore she said unto Abraham, 'cast out this bondwoman and her son.'"

For " He (Isaac) was a man of GENTLE NATURE * * * of devout and blameless life."

PREFACE.

It seems to be very much the fashion in late years for those who have a literary effort to place upon the book mart, to set forth in the preface of these works an alleged valid excuse for "inflicting" the public with the treatise which it precedes, and to request for the errors of omission and commission, which appear in this work, the kind indulgence of the reader, in view of the necessity of the author's addressing said treatise to the existing needs in question.

Such familiar reference to the shortcomings of a composition by its author brings to mind the tradition of the Irish pilot who, while officiating in his capacity in one of the harbors of the Irish channel, declares that there are many dangerous shoals and rocks in this water-way, but, "Faith, he knows them all, ivery wan." Just then, with a sudden shock, the steamship pounds against a sunken rock. "Sure," says the gallant salt, "and there's wan of thim now."

The existing rules and usages of the book interest make it necessary with the reading public that they seek for any literary effort in question. This, together with the ease with which this so-called "infliction" may be cast from those possessed of it, lends much force to the conviction of the writer that, at least as far as this "child of his brain" is concerned, there is in no sense an infliction imposed upon the reader, unless—and this is an afterthought—those momentous and unwelcome truths which appear in this work should produce upon the average Allopathic conscience a positive and abiding infliction.

The reasons for the appearance of this book are chiefly three :

First. The observation of the writer leads him to believe that a large part of the laity of the Homœopathic system of cure have investigated only the *practical* claims of this system, there being no facilities at hand whereby they may acquire an intelligent comparative view of the historical and scientific claims of the two schools of medicine, without wandering through much that bears but little upon what they seek for, and interests them still less. Therefore, do they possess only the convictions based upon their experience concerning the comparative capabilities of these two systems in their attempts to meet and overcome disease. The fact that such are often compelled to sit by and hear the oft rehearsed mockeries repeated, without being able to expose them as they might, seemed in itself to the author to be an adequate reason why this work should be undertaken.

Second. A certain work, by Professor Smythe, entitled " Medical Heresies," which devotes the first ninety-five pages to medical history up to 1790, but sets apart the remainder (more than half) of the book to Hahnemann and Homœopathy has appeared, and since it is, perhaps, the most able of the many constantly appearing attacks upon this school, it certainly deserves our earnest attention.

Third. As the more recent and standard works upon " Medical History " profess to give a history of medicine from the earliest ages, but fail to make any allusion either to the work or existence of Hahnemann or Homœopathy, it appears to the writer that such omission before the reading public should be made good. Consequently, this, the result of the author's humble labors, is tendered to supply the omission, with this proviso,—that these authors, in particular, who are culpable for such defects, do not understand that, by the above statement, we forego or relinquish any of the privileges of copy-right, which our title-leaf certifies to after the duly prescribed form.

Regarding the many lapses which the author has no doubt are to be found in the subsequent pages, the public need not be told who is responsible therefor; the author only wishes that, at this writing, these could be turned to, for they would certainly be most summarily dealt with. It is only in justness to his friend, Mr. Stephen II. Knight, to say, that through his ready aid, many clerical and typographical errors have been expunged, and that thanks are due for this timely service.

As regards his readers, that, in their expenditure in time or money, or both, they will by what they glean from this volume receive an ample equivalent, is the earnest and sincere wish of the author.

SEPTEMBER 10, 1881.

904 FOURTEENTH STREET,

Franklin Park, Washington.

CHAPTER I.

Long before science pointed its first periods, long before the appearance of the rudiments of civilization, farther back into the night of time than the myths of tradition refer, but coeval with those decisive triumphs of man which made the support and perpetuation of the species possible, transpired the birth of medicine.

Different races of men have entered upon their beginning at different epochs of the world's history, and among the most primitive individuals which these races have produced, the appearances of pain or other bodily disorders are experiences as natural as the pangs of hunger or the sensation of bodily cold. Having before them an experience with ailments and suffering it was natural that the effects of various kinds of food upon the system would be noticed, that the influence of the application of certain substances upon the external surface of the body would receive comment, and that these things being remembered, they would invoke their help in relieving and overcoming pain and disease.

From such observations together with the usage

treatment which grew out of an experience in avoiding and encountering physical ills appeared that medicine which was the recourse of the first progenitors of the different races of men.

This medicine was necessarily rude and primitive, for men's facilities of observation were narrow and restricted, and, moreover, a sufficient reason to explain the operose or tardy unfolding of the simplest practices now comprised in the medical art which offer succor to the suffering in their time of need, was the inability of their undeveloped discursive faculty either to understand many of the simplest expressions of the natural laws or to witness the more striking terrene phenomena without acknowledging an extreme fear for what they believed to be a superhuman or miraculous manifestation.

Passing from that era which saw the human family in its first attempts with medical resource, over an interval which covers a stupendous period of time, at length we reach an age which we owe a passing consideration from the fact of its seeing a mother race in the acme of its power and prosperity, a race supposed to be the progenitor of ancient Greece and her European contemporaries, a race flourishing long anterior to these latter nations and passing from being before these nations emerged from the darkness of barbarism or at best were but an uncivilized and pristine people.

So little is certain concerning the history of this Aryan race—their characteristic features yet undetermined, and the question as to what was their place of abode a matter concerning which eminent archæologists disagree—the claims of Europe as well as Asia, to this honor being in turn set forth—that we admit the weight of the prevailing view which avers it only to be fixed as beyond successful dispute that this race appeared, increased, and waxed strong, then waned and passed away to be lost to view forever.

What we have been able to learn of this race does not contribute to the cause of historic medicine, but the Oriental Indians, sons of Iran, supposed descendants of this nation, and inhabitants of that country which lies easterly from Persia to the confines of China, and southerly from the Little Thibet to the Indian Ocean, boasted a high degree of prosperity 3,000 years B. C., attesting the same by ruins, stone memorials, and most of all by that ancient literature, the Sanscrit tongue. In these writings reference to their medicine shows a classification of diseases into specialties, and a knowledge of the existence of the pulse, and the fact of its examination in their methods of judging disease.

The Sanscrit literature was maintained to be the exclusive property of the Brahmins, or priests, this class being the acknowledged authorized dispensers of its benefits, hence to them was relegated the office of physician. They held that their medical knowledge came to them from heaven, and took occasion to consult the stars, the flight of birds, and certain mystic rites, before offering an opinion as to the character or termination of any particular sickness in question. These have been and still are the unchangeable bias of their medical art.

The Chinese people have had a perfected system of chronological data since 2357 B. C.; they exhibit a work upon medicine which lays claim to authorship in the twenty-seventh century B. C.; this work contains most absurd theories concerning the nature, course, and treatment of disease. It has a most elaborate method for examining the pulse, describing its beat at one part of the right arm for a certain disease, and at a different part of the left arm for other complaints; it also directs in this examination the use of several fingers upon the pulse, at one time describing their manipulation after a fashion which we should recognize as being similar to the performance of the fingers of the piano player.

This medical system which has been, and still is, their standard authority claims by this manipulation the power of determining the seat and nature of diseases. This people of paganic worships and practices, of steadfast devotion to their traditions and customs, of uniform deference to their absolute despot, that object of religious awe and worshipful penance, have excluded, until recently, all communication with foreigners this people, though they had produced gunpowder, silk, and the art of printing long before the Christian era, have through all the ages still adhered to the same ideas and practices of the medicine of thousands of years ago, asserting the best treatment for small-pox to be a covering of red for the walls of the sick room and holding the opening of the dead for purposes of information to be a crime, punishable with death.

It has been supposed by good authorities that the Chaldæic or Assyrian people once possessed a greater and earlier civilization than the ancient Egyptian nation; still preserved observations upon their science of astronomy running through an unbroken line of nineteen hundred years, this term of time being completed several hundred years before the era of Jesus Christ, and certain cuneiform inscriptions found upon burnt bricks, and still extant, serve to offer evidence of this nature concerning the Chaldæic Empire.

From everything that can be learned of this people it appears that their priests insisted upon being recognized as the exponents of all knowledge, but their medical practice only consisted of such incantations and magical arts as would excite the superstitions of the ignorant people and secure their reverential respect and servile obedience.

It is with rare interest that we proceed to examine the records and history of a nation, which wonderful in its accumulation of the evidences of a past civilization will, to a more marked degree than that of any other contemporaneous people, be found to further the purposes of our subject.

The earliest knowledge to be gained of the Egyptian nation is from its iconographic treasures, and apart from the question as to which nation may claim the greatest antiquity in its monumental history, we note, that where these chronicles of other races are scanty, fragmentary, and usually vague and obscure, touching what can be learned of their past, the monuments of Egypt—many of them most famous—exist in great numbers, magnificent ruins, statues, columns, obelisks, temples, sphinxes, colossi, pyramids, and the labyrinth, offering one succession of valuable historical aids.

The date of the building of many of these cannot be determined, but to them is conceded a season of great antiquity. Chéops pyramid, the largest of these, has a foundation which covers about eleven acres and a height of about 480 feet, terminating at this altitude in a top about thirteen feet square; this pyramid conforming so exactly with the cardinal points that it casts no shade at noon-day. How many centuries of training in architectural science were necessary to qualify these people to produce such a structure !

The hieroglyphics cut upon these monuments exemplify the Egyptian records, showing them to be the most authentic, the most complete, the most connected, and the most useful to those nations which were destined to appear centuries later, and represent all that was in their day embodied in the progress of the arts and sciences.

Following closely upon this period the practice of papyrus and parchment writing appeared, and henceforth we may rely upon receiving a more complete description regarding the civilization and history of the Egyptian nation.

We learn that they already, to an eminent degree, had perfected themselves in arts, manufactures, and agriculture; by maintaining in their lands, with the aid of systems of irrigating canals, a high grade of cultivation: by displaying in their manufactures great skill in the production of rich dyeing materials, textile fabrics, porcelain, glass; and by being conversant in the arts with the imitation of precious stones after a manner as yet unequalled, of preparing enamel and mastic for mosaics, and, finally, by having acquired such proficiency in architecture as to have conceded to them the most distinguished place among nations to this day.

It will be interesting at this juncture to note the contemporary history of Egyptian medicine. We find one Thoth identified with the first mention of this subject; this personage is described as the inventor of all the arts and sciences, and is also credited with a great many volumes which treat upon a number of subjects, among them, medicine.

The volumes devoted to medicine adhere to a comprehensive plan of classification, one volume being devoted to diseases of the eye, one to diseases of women, one to anatomy, one to instruments, one to general diseases, and one to mendicaments.

No one speaks of having personally inspected the original writings of *Thoth*; historians have been unable to fix even approximately the date of his existence, we are not able to obtain any personal information of him or his antecedents, nor can we learn when or under what circumstances those writings ascribed to him were produced.

But for the efforts of Galen and M. Houdart to prove to the contrary we should speedily charge these works to a rather late period. In view of these efforts, however, we find our most prominent historians pronounced in the opinion that these writings were a production of numbers of philosophers, living in different ages, who, being prevented by a long established custom from exhibiting individuality in their works, prefixed all literary compositions with the common superscription of Thoth, and in such a manner as in our time is used with particular reference to authorship.

From the first that can be learned of the condition of Egyptian medicine it has fallen into the hands of the priests who, declaring it to be a sacred art, assert the right to engross its control. The inferior priests —pastophori, or priest-physicians—under the higher order of priests had charge of the disposition of the sick and indicated the employment of treatment. These priests or priest-physicians were such by virtue of lineal descent, for only the posterity of the priest class could become their successors and be taught their peculiar hieroglyphics and instructed in their peculiar attainments. "He who was born a physician was prohibited equally by heaven and by law from abandoning the occupation of his ancestors."

It is probable that at a very early period it was customary for the sick to be exposed in the public places that those who could speak from previous experience might take opportunity to offer advice. In this way some knowledge of the uses of certain herbs was obtained; this knowledge was appropriated by the priests, and, when employed, was used in conjunction with mystical and ceremonial appeals to their divinities, for the priests insisted that disease appeared at the bidding and because of the wrath of the gods, and hence the need of ceremonial offerings for purposes of conciliation.

Furthermore, the pastophori were confined to the doctrines of Thoth for hints as to the manner of management of the sick, for if disease were met after a method not expressly prescribed in the works of Thoth, should the patient die, the priest responsible for this departure from authorized practices was punished with death.

Thus it will be seen that the established formularies of the priesthood were not liable to be embarrassed by a comparison either with certain indicated improvements or by the legitimate results of efforts at research, for the penalty was sufficient to warrant that those having the exclusive charge of the sick would not care to test the merits of suggestions not sanctioned by this authority.

That these Egyptians under this status were hindered from attempting experimental inquiry into those allied branches of knowledge which go to make the solid ground work of medical science, may be shown by alluding to their ignorance of the first principles of human anatomy. The effects of time upon their dead, victims in wars or other disasters, produced the vanishing of all except the bony parts, and forced upon the Egyptians a familiarity with the tournure of the bones, but with this exception their lack of information in this direction may be shown by referring to their method of embalming the dead, which is described by Heroditus as being performed by first evacuating the contents of the head and body by the use of an iron hook through the nose in the first instance, and by opening the body with a sharp stone in the second instance.

After thoroughly cleaning and washing the body, aromatics and spices were used, then for seventy days the body was subjected to the action of a solution of salt, then being covered with gum, and enveloped with cloth, was finally left in a condition in which it would remain intact through time unlimited. Now the instruments and means of removing the contents of the body is evidence of a lack of knowledge of its component parts.

The superstitious horror with which the people looked upon any attempt to cut or operate upon the dead—the operator upon instituting such procedure was assaulted with missiles and imprecations, used the utmost dispatch while employed, and hastened to éscape immediately upon completing his labors—offers additional evidence in this direction.

Again, we would refer to the fact that the school of

the Asclepiadæ, upon the Greek isle of Cos, was founded many hundred years later than the period to which we are at present giving attention, and that this school, though known to be thoroughly instructed in the lore of the Egyptian priests, yet had but very indefinite ideas of the structure of the human body, during the earlier portion of its existence.

Lastly, we would refer to the views of M. Renouard on this point, who declares that he is justified in asserting that many of the anatomical and other works claimed to have been possessed by the ancient Egyptian priests, were, in reality, the work of some writer of the Alexandrian school since the time of the pursuit of anatomical dissection and research.

The Hebrew nation is closely identified in its earlier history with the Egyptians. These " strangers from the other side," by favor of the Egyptians, occupied a part of the rich country of the Egyptian territory. Certain of the Israelites won favor and prominence both at the Egyptian court and with the learned men of the people. We find Joseph forecasting the days of Egypt's sore trial, and shaping them into a successful and prosperous period. Later on, Moses became affiliated to the royal household, and was thoroughly instructed and trained in the learning of the Egyptian priests. Evidence of his ability while at the Egyptian court is attested by the account of Josephus, which mentions an invasion of Egypt by the Ethiopians, whereupon Moses was made a general, and rendered signal service by defeating the invaders and expelling them from Egyptian territory. Again we find him excelling the wise men in their hieratics, and amazing the priests, as well as the royal family, with his surpassing skill in his manipulations with their magical arts.

Previous to the escape of the Israelites from Egyptian bond-service the intimate relationship and intercourse between these two nations must necessarily have familiarized each with the other's usages and customs, so when we find the Israelites beginning their exodus under the leadership of one so thoroughly conversant with the lore and methods used in Egyptian medicine we are not surprised to find the medical art of the Israelites to be at first nearly like that of the Egyptians, except so far as it might come into collision with the interests and influences of the Hebrew religion.

The knowledge acquired by Moses of the requirements of health and the nature of disease may be partially appreciated by turning to the book of Leviticus, where the laws of life with reference to health are discoursed upon, and the sagacity displayed in marking out the difference between certain skin diseases is manifest.

We soon find, however, that the Levites or priests are following the example of the Egyptian brotherhood, and that they take the charge of medical affairs and associate them with their priestly functions. We may believe that later on these things continued very much after this fashion, for we are told that in Solomon's time there appeared a work said to have been written by this sovereign which taught how to treat disease by natural methods, and which was seized and destroyed by a high priest (Ezechias?) because it would damage the weal of the Levite order and trench upon their ordinances of sacrificial ceremonies.

Lest there may be those who will be disposed to wonder at the considerable mention in these pages of subjects of general historical interest, it is desirable that the pertinency of these matters should be made clear in the following explication of the more prominent corollaries which have been already set forth.

Claiming that medicine entered upon a duration of existence contemporaneously with the first races of men, we have made the effort to indicate the position of medicine in the early states of the ancient nations, to regard its general advancement and to note the highest degree of excellence which it reaches during the progression of the different nations through their many phases of existence.

The grasp of enlightenment manifested, and the high degree of civilization arrived at by certain of these ancient races are subjects of attention, for it becomes necessary that we appreciate these matters if we are to ascertain if the progress of medicine were commensurate with the proficiency developed in the arts, manufactures, and other features of an advancing civilization synchronous with it.

We have read of the skill acquired by the ancient Chinese in the manufacture of gunpowder and of silk; we have read of the wonderful records upon astronomical science which attest how well versed the ancient Assyrians were in astronomy thousands of years before the Christian era; we have read of the marvelous civilization of the ancient Egyptians,

of their comprehensive schemes of agriculture, and
of their accomplishments in the manufactures and
arts that yielded fabrications which to this date encounter no likeness, unless it be in the products of
nature's own laboratory; and we are also familiar
with the mention of enduring examples—master-
pieces of engineering ingenuity and execution—of
which we learn, from the annals of man, that so far
back as those annals can acquaint us these memorials
are still found standing, hoary with age, and offering
to future generations an archetype which none have
attempted to follow, and the processes of the con-
struction of which none have been able to explain.

It must be patent enough to the reader that these
evidences already referred to teem with wonderful
examples of the skill, ability, and genius of these
ancient races, substantiating it beyond peradventure
that these nations must have reached both an enviable point of cultivation and a high degree of intellectual development at this time.

With this understanding of their powers, aptitudes,
and possibilities before us, it seems incongruous indeed that in no case in the history of the ancients
have we seen that the medical art was permitted to
be illumed with the enlightenment and erudition of
2

the day, or that there was incentive offered or effort prosecuted that this medical art might reach a degree of perfection commensurate with the progress made in the several phases of their civilization as just instanced.

On the contrary, the blighting hand of the priestly orders encompassed it within the grasp of its exclusive and jealous care, from which it is not found to take upon itself a different complexion than that which superstitious practices, mystical incantations, or the like, can confer.

The student of the early history of man knows full well that there is no natural quality of the human mind more marked or so eminently inherent as that of superstition or the tendency to regard with superstitious fear and awe even the simplest expressions of nature's forces when their cause or character is not understood. To draw the breath of life is not more natural than this, and it was the recognition of this natural quality that led the priest class to throw the halo of superstition and of magic not only about the investitures of religion but around the medical art as well, with the aim of assuming its control, preferring that their medical skill remain unchanged and unimproved, rather than that the spirit of research be free to open up innovations and new currents of thought, which might be troublesome by conflicting with their dogmas, abridging their powers, and possibly revolutionizing their whole fabric of government, of priestly autocracy and of rule.

However, in the course of the elaboration of the cabal that was the ægis of the priesthood of Egypt, it became necessary that certain important difficulties should be overcome before the complete fulfilment of its perfection of design could commend it, and when in the interest of the priesthood, methods were produced for the achievement of this end, we find in medical and correlative matters that these people are able to offer results which compare favorably with their efforts in the arts and in other vocations.

_ An instance of this is the Egyptian process of embalming already described. Their doctrine of " trial of the dead " held it forth to the inquirer into a future life, that should the tenets of their dogmas be recognized and so acknowledged before the priests, it was the good office of this order to deposit the deceased in the catacombs, there to await *unchanged* the summons to the Elysian fields of Aahlu.

While a plan of catacombs might not be surpass-ingly difficult to project, the need of the procure-ment of a process of preparing the dead so that it might be able to successfully withstand through an indefinite period the assaults which the forces of the seasons would induce, was apparent at this point of the unfolding of their system of faith.

· This called for the production of a master mind, for the undertaking was beset with additional diffi-culties by the vigorous proscription against the ac-quirement of any degree of familiarity with the nature of the subject in question. Hence the methods are found to be both ungainly and prim-itive in what is comprehended as the "Egyptian process of embalming," although this process has presented such marvellous results in the direction desired that we are justified in classifying it among the greatest achievements of this wonderful people.

In conclusion, there are three different points to which we wish briefly to refer:

First. The high capacity and mark of genius which the memorials of these ancient people attest.

Second. That in medicine, and such branches of knowledge as relate to it, the ancients showed as great aptitude and as much natural competency as in other matters.

Third. That the medical art was shackled to mystic practices, and amalgamated to irrational forms of worship by a governing priestly class who were ever ready to construe any matter in question as a threatened encroachment upon their prerogatives or powers, which should be so dealt with by them as to render any weakening or prejudice of their investitures impossible.

CHAPTER II.

It is with fresh interest that we turn from this study of the marked features of the medical art of the most ancient nations to investigate in connection with our subject the general aspects of ancient Greece, a nation which, though at this time in its infancy, was destined, from the pinnacle of its progress in the arts and sciences, to exert an influence unequalled and incalculable both in measure and value upon those races of men which were yet to be.

The first accounts of ancient Greece were preserved at a time when inscriptions or parchment writing were not resorted to or known, and it is believed that at the time of Homer and for a while thereafter no method of writing was used, the minstrels and rhapsodists handing down orally, from one to another, the earliest of the Grecian narratives, in this way preserving them for us.

Through these we are made acquainted with the grand epic poems of Homer, and it was the rhapsodists who carefully reared to future generations the handmaid of literature—Grecian mythology. This authority places the origin of medicine with

Apollo, and refers to the achievements of other of the mythological deities with the powers of healing.

It will be unnecessary to enlarge upon these accounts or to more than allude to the names of Chiron, Æsculapius, Machaon, Podilarius, Achilles, Teucer, Melampus, and Orpheus, who appear most prominently in medical mythology.

It was not unusual for the Greeks—especially during the heroic period—to deify those who in wars or battles displayed personal prowess, and while it is true that extravagant recitals extolled the gifts, and in nearly every instance ascribed the possession of surpassing healing powers to these heroes, it is also true that pronounced or startling medical accomplishments alone were as often the occasion of the deification of the holder.

Æsculapius was said to have been the son of Apollo, by Coronis, daughter of the King of Thessaly. He is accredited with having been the pupil of Chiron, and to first have cultivated medicine, and with such marked success that divine honors were conferred upon him, and thereafter among the Greeks he was styled the God of Physic. The mingled voice of pæan and of adoration rose in his honor, and such fabulous accounts of his wonderful powers were related as the ability to raise the dead—a number of instances being named. This, however, amounted to an indiscretion on his part, for Pluto, the ruler of the lower world, beholding with fear and jealousy this threatened depopulation of his kingdom, determines to clear his domains of this pestilence, and thinking to divorce the doughty Æculapius from all that is included in a partnership with the corporeal member, bears him o'er the Styx, and leaves him within the inner confines of this circling stream, there to reflect upon such offences as must not needs be.

The successors of Æsculapius were his sons Machaon and Podilarius, who acquired great repute from the practice of their profession in the Trojan war. The direct descendants of Æsculapius were accorded the exclusive control of the practice of medicine, and the numerous temples which were consecrated to the God of Physic, and which were the dispensaries where the sick and afflicted came seeking relief, were consigned to their charge. These descendants of Æsculapius were called *Asclepiadæ*.

Historians are pretty generally agreed that certain of the prominent features of the Grecian religion are borrowed from the Egyptians. The fact that Greece presents a situation easy of reach and access by those ancient nations which at an early time were in the zenith of their fame and power, together with the known recurrence of famine and civil wars in Egypt are reasons which would, in themselves, answer for the appearance of Egyptian colonies, and for the engrafting of features of Egyptian customs and doctrines into the Grecian civilization.*

We are interested in this, for we find the medical art in Greece in the possession of an exclusive caste, after the plan of the Egypto-Indian nations, and being handed down from one generation to another, and to only such of their lineage as may satisfactorily pass their secret initiatory tests. The Asclepiadæ in their method of practice had recourse to mysterious incantations and the like, and upon grievous visitations, as the appearance of a pestilence among the soldiers in the Trojan war, superstitious rites and atonements were resorted to. This, with the exception of the recital of cures, which the reci-

---

* Athens is said to have been settled by the Egyptians, and Argos is also claimed to have been colonized by them about the seventeenth century B. C.

pient was under obligation to bring duly inscribed upon a tablet—votive tablet—for exhibition upon the walls of the temple, and the employment of simple dressings and the like upon wounds, comprehended the compass of the art of healing for several ensuing centuries.

We now note the appearance of one who is known as the first philosopher. Pythagoras was born in Samos, about 570 B. C.; his father was the wealthy Mnesarchus. He received instruction at the hands of learned men of the highest repute, after which, according to the prevailing belief of the ancients, he visited, in turn, Egypt, Phœnicia, Babylon, Chaldea, and India.

Pythagoras left no written memorials, so far as is known, and this, together with the fact of the existence of many inaccurate and fabulous statements of his disciples concerning him, renders acquaintance with his history difficult. It is not certainly known if he ever saw the countries above-named, but it seems probable that he lived for a time in Egypt, for he became thoroughly familiar with the mysterious lore of the Egyptian priests. We also find that he asserts and embraces the doctrine of the Hindoo priests of metempsychosis.

Pythagoras was well informed in astronomy, at this time a Chaldaic science, and seems to have been acquainted with what we recognize as the Copurnican theory, which makes the sun the centre of Cosmos. He made also most exhaustive mathematical researches, following in this the Phœnicians and Egyptians in their excelling accomplishment, and he is said to have discovered the proposition that the square on the hypothenuse of a right angled triangle is equal to the sum of the squares on the sides. (Diog. Laërt. VIII, 12.)

He is also said to have paid particular attention to medicine. (Diog. Laërt. VIII, 12, 14, 32.)

Pythagoras is credited with having been a noted athlete, and with having advocated the need of the cultivation of gymnastics and music; also with having held to the necessity of self-restraint, temperance, simplicity in personal attire, silence, virtue, and uprightness in the affairs of daily life. He also is said to have abstained from the use of animal food.

The epithet of first philosopher is applied to him because, on a certain occasion, being asked what he was called, he invented this term, (love of wisdom,) in contradistinction to the title of sophist, (wise,) then in general use.

Upon his return to Samos he rapidly gathered about him a circle of pupils, who showed great respect for "the master."

Pythagoras marks, and was the cause of the appearance of a new era in Grecian intellectual advancement, which had as great an influence upon medicine as upon the other arts and sciences. His methods are seen to be diametrically opposed to those of the Asclepiades, who transmitted their rites only to those of their caste, and enjoined secrecy with these transmissions.

It was the aim of Pythagoras to discuss the primary causes, and to establish, with certainty, human knowledge generally. We can readily understand why he should have been unfavorably received by the ruling caste in Greece, and how, by the tyranny of the ruler Polycrates, he was soon driven from the country. From Samos he went to the Grecian colony of Crotona, in southern Italy, where he established himself. The treatment which he received in Greece probably carried its lesson with it, for we here find him invoking the aid of organization and of politics, teaching his pupils, under the injunction of secrecy, and allying himself and his school with the aristocratic party.

The mysteries of this new order of Pythagoreans, and the strength which this organized union developed did not avail, for we read that his enemies set fire to the building where an immense gathering of his followers was being held, causing the death of a great many of this order.

It is uncertain whether Pythagoras himself perished at this time, or was starved to death soon after, as claimed by some authorities; at any rate, the dissipation of the sect followed.

It will not be in place here to attempt to judge the systems and discoveries of Pythagoras, or to more than refer to his peculiar theory that postulated his philosophy upon numbers as the unit of being, but we wish to give the talents and work of Pythagoras the prominence which they deserve, and to show that though the Pythagoreans were dispersed and persecuted, yet they continued to proclaim the teachings of "the master" throughout Greece with the effect, when the Asclepiadæ found their secrets becoming common property of compelling their priests to lift the veil of secrecy, and to demonstrate their ability to lead in the powers of indoctrination, if they would have their institutions continue to hold their position as centres of learning.

The natural outcome of this change was the free discussion of the principles and practices of the medical art.

It is at this time that a family appeared upon the theatre of medical events, a family whose representatives occupied a· prominent place in the medical profession continuously for three hundred years. Hippocrates I, said to be the fifteenth in descent from Æsculapius, we give but brief mention, but Hippocrates II, one of seven, in a direct line of this name, merits our most earnest attention. He was born about 460 B. C.

It has been previously mentioned that in several localities in Greece temples dedicated to Æsculapius were erected where medicine was separated from other branches of knowledge and made a mysterious art, and after the occurrence of the events just narrated, it became usual to attach medical schools to these temples for the instruction of those seeking to acquire a proficiency in the art.

The isle of Cos possessed a temple and medical school which had become perhaps the most celebrated of any, and it was here that Hippocrates II devoted himself to obtaining a thorough and exhaustive knowledge of what was there taught in these branches. After finishing here he left his own country and traveled in the principal cities of Europe and Asia for the purpose of further perfecting himself in the lore of his profession. After his return he wrote a number of works which, from a medical point of view, rendered the age in which he lived memorable. It is difficult to determine the number or value of his works because of the use of his name by many writers who came after him, but Galen claims to be able to distinguish some of the genuine writings of Hippocrates.

Hippocrates was the author of an important revolution in medicine. The grand principle upon which he stood, was to observe carefully the aspects of nature and to draw his deductions therefrom. In the use of this accumulation of facts which observation had revealed, he sees the operation of universal law and not the influence of good or evil powers.

His descriptions of acute diseases, as we find them at this day even, are wonderfully accurate, and his deductions and comments are often able and practicable. I will here notice the important observation referred to him, to the effect that during the reign of epidemics, the most unlike even of intercurrent affections will take upon themselves a com-

mon likeness; that is, there will be features common
to all. Again he says that nature has an inherent
spontaneous curative power, which is of itself competent to cure disease.

He says, first, every morbid state consists of an
affection and reaction; second, the affection is the
product of a morbific cause; third, the reaction is
the exercise of the *vis medicatrix naturæ*. The medical school of Cos, before his time being the most
prominent one in Greece, became now, through his
efforts, famous indeed. His wonderful skill and
brilliant cures were noised to the farthest corners of
the known world. Cited cures of noted personages
were carefully inscribed and handed down to future
generations.

Unfortunately for him, as it was for several generations succeeding, a knowledge of human anatomy
could not be obtained. The Grecian religion taught
the necessity of the immediate final disposition of
the dead, to free the soul from torment. The prejudices of the priesthood and the people forbade the
examination of the dead, even for the advancement
of anatomical knowledge.

Pythagoras made himself familiar with the
anatomy of animals by dissection; he also studied

3

the anatomy of the human hair, skin, and nails, and we have no reason to think that Hippocrates was possessed of a better anatomical knowledge.

The ideas of Hippocrates concerning the internal structure, composition, and functions of different parts of the human body seem ridiculous to us at this age, and when he attempts to theorize, with this knowledge for his basis, we are not surprised at the outcome. His theory of the four elements and the four humors held that these elements, heat, cold, dryness, and moisture, were the four elements which, combining in varying proportions, constituted all the bodies in nature. That the human body contained blood, phlegm, and two kinds of bile, yellow and black. In health these were in due proportion, a preponderance of one being the appearance of a humor which, in passing out of the economy, would have the effect, as regards the organ enlisted, to leave painful conditions with it.

The theory of *coction* and *crises* held that the morbific material in the economy to be evacuated properly should undergo coction, the totality of the forces of the economy having accomplished this, the symptoms improved, the crisis was at hand, which with the emunctory discharges, terminated the disease in favorable instances. If a sickness terminated unfavorably, it was owing to the inability of nature to elaborate the morbific matter or to produce coction.

It is believed that the prominent physicians of the intervening centuries have, in turn, improved and perfected this theory ; at all events, this theory has had supporters down to the present time.

Of the theory of fluxions his work on the "Regions in Man," says : "Fluxions are also caused by heat because the tissues become rarefied when heated, their pores enlarge, and the humor they contain is attenuated, so that it flows easily when compressed."

. . . "For there are more veins in the upper part of the body than in the lower, and the soft parts of the head are thinner and need less humidity." To these absurd ideas we append the theory of treatment. Holding that the phlegm augments in Winter ; the blood in the Spring ; the bile in the Summer ; and the atrabile in the Fall, the rule for treatment to induce action in the system contrary to that produced by the cause of the diseases, led to the enunciation of the famous therapeutic law *contraria contraribus curantur*. The cause of disease being considered an excess of one of the humors, the one in question was attacked by a remedy with contrary properties.

Hippocrates employed in his treatment blood letting to syncope and fainting, clysters, cataplasms, and suppositories, drastic purgatives, emetics, diuretics, and laid great stress upon dietetics.

In surgery he undertook something more than the simple dressings of the Asclepiadæ; wrote a work upon wounds of the head; in depressed fracture of the skull used the trepan; used the actual cautery; originated a perfected method of bandaging; reduced dislocations and adjusted fractures; and used obstetrical forceps in difficult labor.

We find the prejudices of the age of Hippocrates imposing complete ignorance upon all in certain subjects, and we also find that he lapses into the gross error regarding these subjects of producing *à priori* theories concerning them, theories which are only of interest as exhibiting how preposterous a statement and how false a position even Hippocrates may assume.

In propagating such theories as "the four elements," the "theory of generation," etc., and in instituting "humoral pathology" we find him forgetting his own injunctions to his contemporaries, "*that the nature of man can not be well known without the aid of medical observation, and that nothing should be affirmed concerning that nature until after having acquired a certainty of it by the aid of the senses."*

However, the value of his work cannot be over-estimated. The mystic veil had been torn from the medical art, and his labors have laid an enduring foundation, upon which the imperishable structure of medical science will rise, vast, imposing, and faithful to the demands of an exact and accurate compliance with all the laws of nature.

Though he asserts the law of *contraria*, he further says that diseases are also cured by the similar action of medicine; that is, remedies with properties similar to the morbific cause. The Hippocratic work " On the Places in Man " seeks to illustrate this law of *similia* by stating that the same substances which cause cough, vomiting, stranguary, and diarrhœa will also cure those complaints.

The Hippocratic treatise upon Ancient Medicine declares that remedies sometimes effect a cure not by reason of contrary or similar properties, but that they operate in an inexplicable manner. Hippocrates has been called the father of medicine, and in consideration of these foregoing statements, we can now affirm that he is entitled to this appellation in its broadest sense, and, therefore, those of the profession of every school who follow a course of truth and rectitude, and from a standpoint of proper qualifications are conscientious in their service in the cause of humanity, have an equal right to regard him as the paternity of their profession.

Hippocrates II died when about one hundred and three years of age, and while Socrates his contemporary and Pythagoras before him, each met violent deaths because they were supposed to threaten the welfare of the religious system of their day, we may understand the immunity of Hippocrates from persecution if we remember that it was Pythagoras and his followers that established the right of discussion of the medical art, and while Hippocrates was satisfied with what had been wrested from the priests by the Pythagoreans, and it would have been of Socratic temper to have contested, in the spirit of agitation, the right to explore still other fields, such for example as anatomical research, Hippocrates simply undertook a work that he was permitted to engage in to the extent of his abilities.

CHAPTER III.

While Hippocrates lived, his fame and prestige, together with the spirit of veneration with which his followers regarded his work and teachings, placed him in a position above others so elevated as to render his course and theories unchallengeable, and to give to him a power to direct the course of medical thought, well nigh absolute.

Immediately after his death we find an era of confusion appearing in the medical world. In the school of Cos even, exception to the theories and practices of Hippocrates multiplied on every hand. This experience of the medical faculty in encountering a new and disturbing factor led them to regard the force of sagacity and the wisdom of election which the priesthood had displayed in completely controling and shaping, through an efficient and potent organization, the appearance and course of different methods of medical inquiry, and to believe that the interests and the future of their medical fraternity needed the security of some such protec-

(39)

ting invest as the above sodality had afforded the priest class.

They proceeded, with this purpose in view, to systematize the principal theories and dogmas of Hippocrates. The Dogmatic sect in medicine was the result, an organization in the medical world whose scheme had no intent to command control in governmental affairs, and could not hope to hold the life of an opponent in its grasp as a coercive means of thwarting antagonism to its tenets, as did its archetype, the priest class of the ancients, but which, nevertheless, believed that, from its position of authority in medicine as representing the true and only Hippocratic teaching, this school could not only place the stigma of fraternal unworthiness upon a dissenter, but could, by its affirmation that his accepted and avowed opinions were pernicious and false, place him irrevocably beyond the pale of medical competency, with only disgrace and obloquy to his credit.

Though at different periods we shall find this, the dominant school, employing different methods of coercion as regards its opponents, we can readily understand why this weapon, used at this age, would be of all others the most effective, if we remember that at this epoch the interchange of ideas was by oral methods only.

For the art of printing being not known, and the methods of hand inscription or writing so tedious and slow as to render manuscript making onerous and seldom undertaken except for the purpose of preserving to futurity the opinions and observations of prominent philosophers—the dissemination of knowledge was restricted to such narrow channels as only spoken address* could extend, and these runlets and their sources being under the misdirection of Dogmatism, we find a prominence and unwonted value attached both to the strictures and antipathies of the sectators of this organization.

This school, with its colossal reputation, occupied therefore a vantage ground from which it could successfully withstand all attacks possible to be made upon it under this condition of civilization and of letters; and we shall find this fact so true that in proportion as means are developed for sending forth the erudition of the age in a current with so broad, so constant, and so free a flow that whoever will, may to the fulness of his desire, quaff thereof freely, in that proportion do we find the dogmas and

---

* With exceptions as just noted.

fortified theories of the dominant school assailed, until during the establishment of the Alexandrian library, the antagonism fearlessly takes shape. Under recognized leaders, and with a definite purpose and unity of action, this antagonism at length is able to effect the discomfiture of Dogmatism, the dissolution of its favorite theories, the overthrow of its most celebrated dogmas.

The Dogmatic school, or first medical sect, is generally regarded as having been founded in the first generation after Hippocrates, and when, generations later, the Empiric school took a determinate form in its antagonism to Dogmatism, we find these two sects waging an unremitting conflict with each other, and while on one hand the strength of the Dogmatic school lay even more in being in its tendencies in harmony with the prejudices of the dominant philosophers than in its prestige: on the other hand the prosperity of the Empiric school depended upon the increase and diffusion of knowledge. Alexandria, the intellectual metropolis of the world, offered in its museum facilities for promoting the increase as well as the diffusion of knowledge.

The study of anatomy was here aided by the authorization of dissections of the human body.

Fostered by the sovereign, these methods of research could not but develop new and unlooked for facts, which the Empiric school were ever ready to seize upon to employ in the overthrow of the dogmas and theories of the dominant school. The effects of such assaults may be imagined, and as the pursuit of knowledge could not here be prostituted to the cause of a sect, therefore towards the hand of the foreign invader Dogmatism turned, in the hope that its adversary would be disarmed of this most potent weapon, which had demonstrated a competency to inflict such mortal injury.

When after "the birthplace of modern science" had been turned to ashes by the torch of the invader, and the Empiric school from immaturity of age and purpose was unable to rally from the shock of its disabilities, Dogmatism, from the remnants of its antagonists, sought and obtained the material with which to repair her own disfigured and damaged economy. In other words, the Empiric school, together with the other lesser dissenting organizations which flourished in the days of the Ptolemic library of Alexandria, became extinguished with the fall and destruction of this dynasty, and were merged or embosomed in the dominant sect.

That characteristic feature which we have already recognized as pre-eminently peculiar to the Dogmatic school, of assuming an inherent right to declare what is the true and only exposition of all new or unsettled subjects which appear upon the medical horizon, we notice still presenting itself as prominently as ever, as indeed it will be found to do throughout the entire future history of this school. We shall see that the exercise, by Dogmatism, of this aptitude of putting forth the plenary and orthodox ultimatum relative to matters of medical theory and treatment, which the discoveries of science presently demonstrating to be untenable and dangerous, gives rise, particularly in later times, to these arraignments, in which the most prominent scientific observers stigmatized, in the most emphatic terms, the blunders and incertitudes of the ruling school in medicine.*

---

* That the proper significance may be attached to the above, we append a few extracts from the writings of some of the most famous among the scientific and learned men of their time, and who pass judgment upon the merits of the system of the dominant school. The renowned and philosophical Bichat, whose researches and discoveries in anatomy and physiology have placed him in a most enviable position in the profession says : " To what errors have not mankind been led in the employment and denomination of medicines. * * * The same identical remedies have

The sons and son-in-law of Hippocrates, Thes-
salius, Draco, and Polybius, were the founders of
the Dogmatic school and the successors of Hippocrates. These, with Diocles and Praxagoras (of whom
there is some mention, though only very vague and
unsatisfactory accounts,) were the last of the Asclepiadæ. It was, at this period, considered necessary
among the erudite to acquire a knowledge of medicine as well as of the other branches of learning,
and hence we often find the prominent philosophers
of the day expounding upon and referring to medi-

been employed under different names according to the manner in
which they were *supposed* to act. * * * So true it is that the
mind of man gropes in the dark when guided only by the wild-
ness of opinions. * * * Hence the vagueness and uncertainty
our science presents to-day—an incoherent assemblage of incoherent opinions ; it is of all the physiological sciences that which
best shows the caprice of the human mind. What do I say ! It
is not a science for a methodical mind, it is a shapeless assemblage
of inaccurate ideas, of observations often puerile, of deceptive
remedies and formulas as fantastically conceived as they are
tediously arranged."

John Abercrombie, a celebrated physician and surgeon, ranking
the first consulting physician of Scotland, in his time, (first half
of the nineteenth century,) declares that "medicine has been
called by philosophers the art of conjecturing—the science of
guessing."

Sir Astley Cooper, F. R. S., an eminent English surgeon and
professor of anatomy, author of several standard medical and
surgical treatises, and surgeon to the king, exclaims : "The science
of medicine was founded on conjecture and improved by murder."

cine with the air of a familiar. Indeed, we find
certain philosophers, partly because of their indoctrinations and matchless repute in letters, becoming
authority in medicine to the extent that their expo-
sitions of abstract medical questions are accepted
because of the weight which their authorship at-
taches. Plato, and afterward Aristotle, became
ardent supporters of Dogmatism, and brought valuable help to the aid of this school in counteracting
the effect of the opposition of the adverse sects.

Plato's philosophy of synthetical and intuitive

From the great Francis J. V. Broussais, a name familiar to
every student of physiology and of medicine, we quote the assertion, that "when I would seek a guide among the authors es-
teemed the most illustrious and to whom medicine confesses itself
the most indebted, I found nothing but confusion, all was, so to
speak, mere conjecture."

Dr. J. Mason Good universally conceded to have been one of
the most prominent physicians, accomplished linguists, and note-
worthy authors of his day, exclaims that "the science of medicine is a miserable jargon, and the effects of our medicines upon
the human system in the highest degree uncertain, except that
they have already destroyed more lives than war, pestilence, and
famine combined."

Lastly we add the evidence submitted by Dr. Oliver Wendell
Holmes, professor of anatomy, and a brilliant and prolific author
and poet of the nineteenth century, who tells us that "with the
exception of morphine and sulphuric ether, I firmly believe that
if the whole materia medica could be sunk to the bottom of the
sea, it would be all the better for mankind and all the worse for
the fishes."

methods tested the truth of all propositions or discoveries by mental intuition solely, without the aid
and entirely independent of phenomenal evidence,
hence we may expect to find him to be grossly in
error when he attempts to explain the nature and
causes of phenomena. for we see him depending
upon imaginary analogies in evolving his assumptions, the absurdity of some of which may be
instanced by referring to his descriptions of the
arrangement and location of certain of the parts of
the human body.

Aristotle, his pupil, who pursued different and
better methods of philosophical investigation, acquired, under more favorable circumstances, even a
greater ascendency over the minds of the erudite of
his day as well as over those that followed him during the succeeding centuries. These philosophers,
though not, properly speaking, attached to the medical profession, yet were in their time the exponents
of the Dogmatic school in medicine.

In an assumption of Aristotle's we find a legacy
bequeathed to Dogmatism, (and of necessity therefore its accepted doctrine.) This assumption de-
scribes a means of communication between the
heart and trachea that provides for the direct pas-

sage of a current of inhaled air into the cavity of
the heart. The harmful influence of this absurd
doctrine upon the treatment of many diseases may
be imagined.

The wonderful conquests of the Macedonian campaigns under Alexander lent an extraordinary
impetus to the intellectual activity of the time, and
the results of these campaigns extended facilities for
the acquisition of knowledge, by which the Alexandrian library and museum were materially aided in
the establishment of the "mathematical and practical schools of Alexandria, the origin of science."
The founding of these schools, and the successful
organization of the other institutes which should go
to make Alexandria the great centre of letters and
of science, drew the philosophers and students from
every quarter of the known world, and we shall,
henceforth, during Alexandria's glory, find the most
prominent Greek philosophers identified with her
and her schools, and giving the world the benefit of
their labors as residents of Alexandria and devotees
at her shrine of learning.

It is about this period that we find the disruption
in the profession, producing distinct and opposing
sects in medicine. The Empiric sect at Alexandria

was, according to Celsus, founded by Serapion, though other authorities fix it upon his teacher, Philetius. It will be noted that after the dissolution of this school at Alexandria, its past history reposed with its opponents, and of the very numerous writings of which there is mention, and which we have every reason to believe were made by many prominent physicians of this school, none have been discovered, and it is generally conceded that, without exception, all have been destroyed; hence it is from the comments of its opponents that we may learn of its organization, doctrines, and work.

The sects, as opposed to each other, were called Dogmatists and Rationalists, or Empirics. The Empirics declined to subscribe to any of the Dogmatic doctrines or postulates. We have already attentively followed the assumptions of the Dogmatic school as regards what it claimed for its dogmas and edicts upon all problems in medicine, settled or unsettled, and, by its assumptions of what shall be the accepted views of medical matters, how it claimed, from these postulates, to arrange and direct the proper medical treatment to be employed.

The Empirics let no opportunity pass to oppose all this, preferring to commence at the beginning,

4

and from beneath the accumulated débris of centuries of Dogmatic falsity, to disinter the truth, and,
under its immaculate insignia, to observe and cultivate those methods of treatment which experience
would prove to be the most useful and beneficial to
afflicted mankind.

Serapion attacked the tenets of Dogmatism with
much power and vehemence, denouncing the
theory of the essence of disease as absurd and false,
and the assumption of the theory of prevailing
humors, etc., together with the law of *contraria*, was
exclaimed against in emphatic terms.

The most characteristic feature of Dogmatism, was
to resort to the attempt to solve medical phenomena
by recourse to recondite and dreamy theories; on
the other hand the Empirics professed to exclude all
theories from their system, preferring to be guided
solely by the minute description of the symptoms of
a disease, depending upon the personal observation
or related account coming under the careful observation of unexceptionable authority. The name of
a disease had but little weight with them, the total
assemblage of symptoms was carefully noted, and
such treatment exhibited as had been found to be
most successful against such a concourse of symp-

toms. Any given set of symptoms with the successful methods used to overcome them, was termed an
Empiric theorem, and each Empiric physician endeavored to acquire as many theorems as may be.

Thus it will be seen that the Empiric school, by
imposing fixed and well-defined rules for proceeding,
in observing and estimating disease, and in persistently opposing the false and unsettled theories which
were a basis of the treatment of the Dogmatists, conferred upon the medical art and upon man an invaluable service.

It is probable that the arguments of Greek Pyr-
rhonean skepticism offered encouragement and suggestions which were of value to the Empirics in
refuting the definitions of the Dogmatists, for the
Empirics declined to accept or consider any proposition which could not be submitted for examination to the senses. The Empirics endeavored to
avoid the error, into which the Dogmatists had
fallen, of searching for and assuming to be familiar
with the essential cause of disease; in so doing, they
rejected all theories from their system, answering
inquiries into abstract questions by declaring it to
be impossible to explain what is inappreciable to the
senses, and, therefore, it must be wrong to perplex

oneself with inquiry into occult causes. This induced among the Empirics an indifference amount-
ing, in such matters, to apathy, and we find Galen
authority for the statement, that in the pursuit of
anatomical and physiological knowledge, the Empiric school made no progress.

In other words, the teachings of the Empiric school
would enjoin upon its followers the uselessness of attempting to discover the underlying causes which
produced certain chains of symptoms; hence we see
they were prevented from simplifying their methods,
the result being that their record of symptoms multiplied to such an extent as to produce dire confusion. Attempts at arranging these into species, or
of placing those of a genus together, did not result
in obviating the difficulty, and this weak side in the
Empiric system led to the appearance of another
sect, the Methodists.

The great object of this sect was to simplify the
practice of medicine, Asclepiades, its founder, claiming that all the tissues of the body were pierced
with minute pores, and that disease was simply a
condition of contraction or expansion of these pores,
he denounced the heroic treatment of Herophilus,
and substituted the employment of friction and

physical exercise, and placed great stress upon diet. This theory was amplified by later Methodists, but, owing to a lack of harmony among themselves as to what diseases constricted and what dilated these pores, it finally became remodeled to the extent that two new classes of disease appeared in their category, to divide the honors with the two classes above mentioned.

The Methodists, or "Solidists," as they are also phrased from their belief that disease resided in the tissues and from their opposition to the "Humoral" theory, that of disease being occasioned by some prevailing blood humor, as the Dogmatists assumed, did not cultivate the study of anatomy and physiology to so great a degree as did the Empirics, and we find them as essentially dogmatic in their assumption of theory as the Dogmatists themselves.

An offshoot of the Methodists were the Pneumatics, and still another sect, the Episynthetics or Eclectics, appeared. These latter, in the bewilderment of the many presenting theories and systems of medicine, declined to adhere to any fixed principle for guidance or belief, but left each practitioner to choose a few materials from each of the different systems with which he could erect to himself an edifice, the com-

ponent parts of which, coming from these several
sources and intended for very diverse uses and necessarily differing extremely in their character and
nature, would apparently produce a structure as
heterogeneous as the most eclectic taste might require.

For more than three centuries, just preceding
the Christian era, Alexandria was the home and
centre of science and literature. During this time
science, art, and the liberty of criticism were fostered
and protected by the ruling Egyptian dynasty. The
medical art progressed here after a manner which
had seen no parallel in history, the controversies and
disputations between the differing medical schools
having the effect to bring into especial prominence
the exclusive advantages and conceits of each system.

This " divine school of Alexandria " was the only
place where, previous to and during this period, the
dissection of the human body was permitted. Here
dissection and vivisection were not only permitted,
but facilities for unimpeded investigation after anatomical and physiological facts were offered. We,
therefore, may expect to find that great advancement
was here made.

The physicians identified with Alexandria and

her institutions, and enjoying her unexceptional facilities for the promotion of discoveries in anatomical and physiological knowledge, frequently immortalized themselves while in the pursuit of their chosen branches of research, Erasistrates and Herophilus being, perhaps, the most prominent in this particular; the latter, it is said, dissected upwards of six hundred human bodies during his investigation into anatomical science.

Concerning the future of Alexandria, we find that certain political events arose and cut short her military authority and influence, which was, however, in extent of time, not great. However, from a point of view which shall give due appreciation to her intellectual achievements, without enlarging upon her progress in the different branches of knowledge, only referring to such names as Euclid, Archimedes, Ctesibus, Hero, Apollonius, Hipparchus, and Ptolemy as connected with her schools and the discoveries made in them—aside from the unhappy effect, extending down through intervening centuries to this present time and arising from the intellectual misdirection here instituted by Aristotle and the Dogmatic sect in medicine, which misdirection perverted the influence of Alexandria in the near future, and

thereafter from the true aims of scientific research
as applied to medicine but which insured the employment of *à priori* methods in her supposed
cause—aside from this, indeed, we may affirm that
the adherence to those simple and truly philosophical methods of investigating nature and her laws,
which, under Alexandria's fostering care, were subject to no check or limitation as regards the direction or scope of proposed inquiry, but attended, as
they were, with such wonderful results as we have
already noted, invested her with a hitherto unex-
ampled measure of authority and influence in the
realm of science, and will assure to her, as the intervening eighteen centuries already bear witness, a
future authority and influence which shall be
strengthened and augmented in proportion as that
science progresses and develops which she originated,
and during the darkness of prejudice and ignorance
carefully cherished and propagated.

The overthrow and desolation of the ruling
dynasty of the time was attended with the extinguishment of this spirit of scientific research. The
different schools of medicine were therefore affected
to such extent as to exhibit a condition of complete
apathy.

After the vanishing from view of the Ptolemic dynastics, had the general state of enlightenment been such as to appreciate and patronize the purposes and efforts of the Empiric school, what this school had already achieved would have marked but the conclusion of the first period of its mission, in which event we should have seen all the opposing sects of Dogmatism attracted to its standard, and reuniting them under its leadership, in the ardor which attends new issues, the Empirical school then would have effected, with that success which characterized its past, the utter demolition of other of the false positions of Dogmatism. However, the intellectual darkness which followed the destruction of Alexandria, was destined to cover all the scientific lore so carefully nurtured and cultivated there, lying fallow through many centuries to come.

With the exception of the Dogmatic, these schools of medicine could not seem to rally from their paralytic state, and we shall readily understand why, if we consider, as has already been set forth in these pages, that the salient traits of Dogmatism were agreeable to the prejudices of the dominant philosophers; for from their station as expounders of wisdom there was something alluring in the *à priori* methods of invoking the imaginative faculties to their aid in shielding their prejudices while interpreting facts or unfolding theories. This, together with the later feature of Dogmatism—its intolerancy of opposition in matters which implicated the fundaments of the system—was an autocracy which the schemes of Dogmatism assumed, and under the shelter of which her exponents might repose and feel an assured protection from any danger of the thorough exposure of their errors, which an unrestrained discussion would produce.

These characteristics to which were attached increased weight and power as being the true methods and expedients of Aristotle, were indispensable to Dogmatism, that hydra which exhibits from its uncauterized wounds received at the hands of the Empirici renewed development and superadded powers and fastening its tentacles about the prostrate forms and institutes of the medical progress of the past, extracts from them their combined roborant qualities, by which, in addition to its own inherent capacities, it is enabled to display a degree of vitality upon which the continued attrition of two thousand years will produce no perceptible enervating effect.

From this time we find but little attention bestowed

upon either dissection of the human body or practical research. The undisturbed occupation of the medical arena by the Dogmatists gave the *à priori* methods the exclusive charge of anatomical and physiological topics and we find this general condition of medical inquiry to continue through many ensuing centuries.

Celsus Aulus Cornelius, a Roman of whom there is much doubt concerning the time at which he lived, his origin, or even his profession, wrote in an admirably pure, concise, and elegant style many treatises, remarkable for their force and ability, upon a wide range of subjects. Of these, such subjects as rhetoric, philosophy, jurisprudence, war, agriculture, and medicine have attracted attention, though at present excepting his work upon medicine and a few others of his scattered writings, none are extant. This work, "De Medicina," consisting of eight volumes, has come down to us, and was for many centuries widely noticed. It is generally accepted that Celsus appeared upon the theatre of medical events in the second century of the Christian era, and it is probable that he was a contemporary of Galen, for we find that the latter does not refer to him in his own writings.

Although we have some accounts of the surgical as well as the medical methods of Celsus, it is generally thought that he devoted himself in medicine to literary work, aiming to distinguish himself in his compilations and history by a purity and elegance of diction, and an impartial and correct depiction of his subject.

Galen, of whom we have just made mention, was a man of wonderful genius and extraordinary acquirements. Born in Pergamos, about the year 131 A. D., he had, in his early youth, the advantage of the tuition of his father, an architect by profession, and also well versed in the erudition of the time. Afterwards being placed by his father under the tutorship of those professors most celebrated in their several provinces of learning, he became thoroughly conversant with such branches of knowledge as philosophy, mathematics, and dietetics. Finally, however, he decided to devote himself exclusively to medicine and its offices. He applied himself to this art, pursuing an equally comprehensive plan of study, traveling in distant countries, and attaching himself to various schools for the sake of acquiring information.

It is credibly stated that one of his longest sojourns was at Alexandria, whither he betook himself to become more familiar with anatomy. Without doubt, his talents and accomplishments place him as the most eminent physician of his age and school, worthy of being ranked as the peer of Aristotle. This brilliant man was the author of a great many (several hundred) treatises upon medicine, and was the first who gave to literature and the profession a complete system of medicine.

Upon finishing his education in medicine, he repaired to his native city, Pergamos, to engage in the practice of his profession, but a revolt breaking out there, he left the city, and went to Rome, where he acquired a widespread renown as a skillful and brilliant physician. Upon the appearance of a pestilence of an epidemic character in Rome, he embarked therefrom for Greece, returning soon after, however, at the instance of the Roman Emperor.

Now that the Empiric school and its coadjutors were no more, and an indorsement of the principles for which they had fought against Dogmatism would not revert to their benefit, we find the exponents of Dogmatism tacitly acknowledging a consciousness of those vulnerable points of their school and against which the Empirici had directed its most successful assaults, by showing an inclination to avail themselves of the equipments which the Empirici had

worn and so long identified with their methods, and
also by a desire to appropriate any degree of eclât
which the sole use and custody of these equipments
might attach.

Celsus and Galen are each amenable to this indictment. Though Celsus displays in his attachment
to "Hippocratean methods," no less than by his
endeavors to adapt his observations to his theoretical
concepts, his allegiance to the Dogmatic school, (for
example, where he accepts the doctrine of coction of
humors and with such unfortunate results,) yet we
see him declaring in favor of certain of the views of
the eclectics, and at another time testifying to certain
excellencies in the peculiar practices of the Methodists.

Galen, in more exact phrase, declares himself to
be attached to no sect, but to study and use from
all, and if we knew nothing more than this of him
we should class him and his statements as being dis-
tinctly eclectic. We may believe, however, that this
sentiment of his is pronounced for effect only, his
practice and utterances stamping him most conclusively as a strong adherent of Dogmatism. He
accepts the Dogmatic theory of the four elements,
which teaches that all changes in the body occur

because of the action of those elements; he also creates a division of eight kinds of temperaments in which these elements are compounded in different proportions; he also indorses Aristotle's theory concerning the human soul, which holds that the soul is divided into three parts; one of which is vegetative and resides in the liver; one irascible and resides in the heart; and one, being rational, dwells in the brain.

From such theoretical bases for his starting point, and without examination or corroboration of presenting conditions, he does not hesitate to declare the rationale of all the phenomena of the economy.

He declares his colleagues in medicine to be "ignorant or punctilious dialecticians, whose discussions are frequently repugnant to the most common sense."

The insignia of dogmatism, "*contraria contraribus curantur*," he uplifts as the distinctive symbol of the school, and to the assumption, of his predecessors subjoins his own subtleties of reasoning, presenting, therefore, as we see, " Dogmatism, amplified, explained, and pushed to its last consequences."

Galen and Aristotle may be justly represented as the equally supporting pillars of the edifice of Dogmatism.

Galen had, as we may now appreciate, such a passion for theorizing that this preoccupation prevented his applying himself to the accurate description of disease in which Hippocrates so wonderfully excelled. He prided himself upon his skill in prognosis, and does not show much taste for surgery; his brilliancy and genius, as exhibited in his many works which treat upon medical subjects, conceded to him more than to any other physician before him the position in the Dogmatic school of being not only its exponent, but its *oracle*, the dicta of the master was accepted because of its source, and became not a matter for discussion, but a belief, a faith.

Without undervaluing the talents and learned accomplishments of Galen, it is nevertheless true that his appearance was a misfortune for the medical art; not in view of the possibilities or the benefits which his talents might have accomplished in other directions in the cause of mankind, but because the scheme of Dogmatism perfected in the Galenic mould, at one fell blow, cut short his as well as its own limited usefulness; for its spirit, grasp, and control of the medical world by indicating Galen as its *locum tenens*, gave to those hypotheses and subtleties, which Galen furnished in explanation of the phenomena and functions of the economy, a weight and authority intended to be infallible in its character.

Physicians, therefore, found the interrogation of nature, in the pursuit of the study of medicine, neither profitable nor inspiring when the dicta of Galen was constantly disputed in the prosecution of original fields of research; consequently, we shall find the medical art through the next sixteen hundred years to be securely bound hand and foot by the tethers of the Galenic school of Dogmatic medicine.

With the disappearance of the medical schools into the vortex of Dogmatism ; with the disposition of charlatans (or those whose methods are deceit and artifice) and quacks (or the uneducated and grossly ignorant) to appropriate the remaining vestige of the Empirici—its good name and prestige—with the result that they ultimately bring it and themselves into degradation and disrepute ; with only a fragment or form of the Alexandrian school in existence from the halcyon days of Rome down through its degeneracy ; with no comprehensive library or educational centre extant, which could to any appre-

ciable extent take the place of the first or Ptolemic
library and museum of Alexandria; and with the
Dogmatists in the exclusive and complete control of
medical progress and development, in the continued
tenure of which they desired to adopt only such
features of the Alexandrian methods as the trans-
planting of its system of slave clerical labor for
manuscript making, with intent to produce and
preserve what would prosper the cause of Dogmatism, we see the scheme of the Dogmatic school,
with all that it implies, practically unopposed, complete and undisputed in its sway in the province of
medical science and thought.

As a consequence those circumstances and conditions which would result from the fullest exercise
of the purposes and influences of Dogmatism would
appear and choke all else in their domain by the
press of their rankness and exuberance. An example of the results of the influence of Dogmatism
under these, its most favorable circumstances, is
shown in the servile condition of the profession; for
freedom of thought and inquiry had been abridged
and estopped until, in the nature of things, the profession had acquired an abjectness of disposition
which suspended the progress of medical science,

until, from the time of Galen, we find the profession with folded hands, passive and satisfied, displaying no disposition to take up the cause of medical progress and further its promotion. Indeed, the medical writers of the Christian world from the time of Galen, for more than a thousand years, were, in the language of Sprengel, but "frigid compilers or blind Empirics, or feeble imitators of the physician of Pergamos."

The gloom which fell upon christendom, and enveloped it during the next thousand years, precluded the interposition of the helping hand of the correlative sciences, which so often, under other circumstances, in the discovery and propagation of new truths, had led and stimulated medicine to new and better directed efforts in the cause of her own progress.

So much need, at this age, had medicine to depend upon her own gropings that we find the priesthood subverting their own trusts and assuming and directing the care of disease, employing expiatory offerings, sacred ointments, the repeating of talismans, the burning of amulets, the wearing of sacred and blessed charms, declaring such methods not only suitable and eminently adequate, but also

the only legitimate and proper relief for the suffering to look to, asserting moreover that, to obtain
alleviation or relief from suffering by other methods
than those directed by the church and her representatives, would debar the offender from the
function of her beneficent offices, and dispossess
him of that staff upon which he needs must lean
through the valley and shadow and up the celestial
mountain.

With medicine under the complete autocracy of
Dogmatism, and with such external influences as the
above to affect it, we may, from the foregoing, fully
appreciate its degraded and stationary condition in
the Christian world from the time of Galen to the
fifteenth century; the only illuminate rays which
fell upon this art here being such Galenic "compilers" as Oribasius of the fourth, Aëtius of the
fifth, Alexandria, Trallianus, and Paulus of the
sixth century.

From this time, through the dark ages to the
fifteenth century, the gloom thickens, when sudden
developments in the civil and political situation of
the Christian world are found to bring about new
and unlooked for changes in the condition of science
and of literature.

CHAPTER IV.

THE DESCENDANTS OF ISHMAEL PROTECT AND NURTURE THE NAMESAKE.

At this period, when the European countries were shrouded in the darkness of ignorance, a sudden and unlooked for change was wrought in the condition of an original and unmixed race, whose home was a peninsula in the southwestern extremity of Asia, to a great extent an arid and barren desert. The Arabs, proud and wild in their freedom and independence, with "their hand against every man and every man's hand against them," were the only people of antiquity who could boast of never having suffered the invasion of the foreign conqueror to prevail.

One appearing among them, and proclaiming himself a prophet of the living God, succeeds in inducing this nomadic people to renounce their paganic religion and practices and to espouse and acknowledge the new religion preached by Mahomet, the Prophet. The endorsement of Mahomet's claim to the power of revelation by his followers helped

to make it appear that his utterances had the sanction of divine authority, and these, with his writings,
were collected to make the Mahometan Bible or
Koran.

The firm belief of his disciples in his power to
perform miracles; his reputed visit to heaven, where
he is alleged to have talked with Moses, the prophets,
and even Deity himself; the talismanic cry of "There
is but one God and Mahomet is his Prophet;" his
simple, abstemious, earnest life, consecrated to God
and his fellowmen; the religious system which he
founded, so peculiarly well adapted to the nature
and race characteristics of his people—being a half
religious and half military system—were each cal-
culated to aid to a considerable extent in making
the future great undertakings of this race possible.
Mahomet, before his death, at about fifty-one years
of age, saw the firm establishment of the new religion upon an enduring basis, but he did not leave
behind an organized government which could weld
and hold together those powerful factions united
under a common religion. Mahomet's martial chiefs,
however, effected a coalition under the leadership of
the father of a wife of Mahomet, who was styled
lieutenant or caliph.

The Mahometan religion enjoined frequent ablu-
lutions, prayers, fasting and almsgiving; sanctioned
polygamy; taught the unity of God and, concerning
a future life, the doctrine of predestination, saying of
this " No man can anticipate or postpone his predetermined end. From the beginning God has settled
the place in which each man shall die." The im-
plicit acceptance of this, and the added conviction
that it was the duty of all Mussulmans to aid in the
diffusion of Islam, and to prosecute its acceptance,
if need be, by fire and sword, were competent, we
may believe, to incite the religious ardor and enthu-
siasm, which this faith must produce upon this
warlike people.

With this understanding of their convictions and
nature the accounts which history gives us of their
wonderful conquests and exhibitions of personal
bravery may be read with much interest and appreciation. Under their first Caliph, Abu Bekir, they
entered upon an era of foreign conquest. The Per-
sian Empire was invaded and finally conquered.
The subjugation of Syria was effected and the cities
of Damascus, Emessa, and Jerusalem, were reduced
by these victorious Saracens.

The whole of Asia Minor being at the mercy of

the Mahometans, the Caliphs turned their eyes toward
the west, and Amrou, one of their generals, who
was sent against Egypt, subdued this country with
its cities, Memphis and Alexandria. The words
ascribed to the Caliph Omar: "If these writings
agree with the Koran they are useless and need not be
preserved; if they disagree, they are pernicious and
should be destroyed," were said to have referred to
the library then at Alexandria, and to have been the
reason of the destruction of this later collection.

From this point the Saracens gradually extended
their conquests over the whole of northern Africa,
conquering and absorbing the Moors, who adopted
the language, name, religion, and customs of their
Arabian masters. Again pushing onward they
crossed the straits of Hercules, took Mt. Calpe,
(Gibraltar,) invading Spain, and ultimately taking
possession of most of her towns and citadels.

After these feats of arms the Mussulmans again
looked northward and beyond the Pyrenees, and
again taking up the sword of foreign invasion they
penetrated to the centre of France. At this moment,
the Franks thoroughly alarmed ceased their inter-
nicine warfare and united to do battle to the common
enemy. Under the leadership of Charles Martel, upon

the plains of Poictiers the Saracen army was defeated and scattered with terrible slaughter, being shortly afterward compelled to relinquish their hold upon a large part of their European conquests.

With this allusion to the rise and triumphs of Mahometanism we shall be able to appreciate the influences which gave rise to their subsequent founding and promoting of science and learning in the East.

The time came as a result of the extensive conquests of the Mussulmans when an intellectual stimulus was engendered among these conquerors, and when, after the time of Omar, others succeeded to the Caliphate who had different and more liberal views than he or his predecessors, their religion, before, an anthropomorphic one, became, with these later and more philosophical followers of Mahomet, a religion whose God with his attributes they contended it was impossible to determine or know of through a knowledge of the attributes of man. Such philosophical features of Mahometanism operating as an incentive, to many of the more advanced Mussulmans, resulted in their attempting to establish for the glory of Mahometanism a celebrated and unequalled centre of scientific and literary knowledge,

which should example, to such an extent as her
triumphs of conquest had done, the power and fame
of Islam.

Omar was assassinated about 656 A. D., and about
one hundred years later Caliph Almansus founded
for the capital of the Empire of the Caliphs the city
of Bagdad, which situated upon the banks of the
Tigris, sixty miles north of Babylon, became, under
his fostering care, a magnificent city of picturesque
groves, palm trees, minarets, and mosques. It is
said to have finally reached a population of two
millions of inhabitants and to have become the great
centre of commerce and wealth.

The caliph showed great zeal in protecting and
encouraging letters. The philosophers of Alexandria dispersed at the destruction of the library
and museum, as well as those who were obliged
to flee from christendom as heretics, were attracted
to Bagdad by the generous spirit displayed by
this caliph toward all, who without distinction
of country or religion, were by reason of their
talents and erudition, able to assist in establishing
a treasury of scientific and literary knowledge in
this capital. A large number of scholars devoted
themselves to translating the lore of Greece into

Arabic, and the writings of Aristotle, Euclid, and Galen, received especial prominence, and were, in fact, the basis and beginning of Arabian letters. Soon the diffusion of knowledge, which its systematic encouragement effected, occasioned the establishment of schools, colleges, mosques, and hospitals, together with libraries, which in Bagdad, Damascus, Alexandria, and Cordova, (in Spain,) reached notable proportions.

These Mussulmans exhibited a love of splendor which was displayed in their cultivation of decorative painting and music ; in science they introduced the system of Arabian numerals, algebra, and astronomy into Western Europe. Science and learning with this foothold, and under later caliphs, was even more ardently patronized, and took rapid strides in progress. Medicine received distinguished attention, and was prosecuted with vigor and success.

It was probably the second century of the hegira that saw the first gathering of the erudite under the patronage of the Arabian dynasty. These philosophers, from a disposition to follow Galen and the Galenic methods, were culpable to the extent that they confirmed certain of the prejudices of the Mussulmans ; therefore, the further development of

anatomical and physiological knowledge was inter-
fered with and interrupted by the prohibition of
dissection and like investigations.

In place of such research and inquiry these philosophers evinced an inclination to devote themselves
solely to translating and delineating upon the tenets
of Galen in declaring their position upon any medical topic.   From this time until the eleventh century
we find the names of many earnest and able toilers
within the belongings of Arabic medicine, of these
Avicenna, though pronounced the most talented and
gifted of them all, and sometimes called the "Prince
of Physicians," is a fair type or representative of his
associates in Arabic medicine.   Avicenna early
evinced a great thirst for knowledge and devoted
himself to a thorough and exhaustive study of the
writings of Aristotle, later he turned his attention to
medicine, and became an ardent admirer of Galen,
whom he follows implicitly in his methods and bias,
even surpassing them in his speculative efforts to
pronounce upon the nature and treatment of disease
depending in regard to the latter, in true Galenic
fashion, upon a preconceived hypothesis rather than
upon actual observation, believing that the first duty
of the physician was to defend the tenets of Galen
from any and all objectors.

With the Arabic physicians, however, there appear three distinguishing marks:

First. Though a servile following of Galen was early established, the imperfect translations of his writings being plenary in their supposed ultimate dicta, yet merit and the result of research, within those limits not expressly forbidden, was, by those Arabic protectors of science and letters, fully recognized; and, therefore, we see that, in certain directions, valuable accessions were made to the general store of medical knowledge by accurate and minute descriptions of diseases, Rhazes being first credited with having written of the small-pox and measles.

Second. By their explorations into the vegetable kingdom, the materia medica became greatly enriched, such herbs as rhubarb, jalap, looch, cassia, senna, manna, and many aromatics being thus introduced.

Third. The Arabian art of alchemy was a contingent which contributed very materially to Arabic medicine. Alchemy, springing from a belief held at this time that the other metals could be transmuted into gold, were but the correct method of procedure understood, led to much secret effort to discover the process of transmutation. Thus, at

length, appeared a new class, the Alchemists, whose
ignis fatuus still led them on, unrewarded as regarded their absorbing aim, yet, though each alchemist zealously guarded his labors in secret by using
in his memoranda cabalistic characters and secret
signs, their incidental discoveries of new chemical
combinations, processes, and substances, served to
prosper, not only medicine, but the several arts and
sciences then extant. Presently they manifested
more definite ideas of their subject; and notions,
which smacked strongly of a familiarity with the
teachings of Aristotle, came to view.

They regarded gold as having a characteristic
"form" separable from its gross part, by which, if it
could be isolated and controled, the alchemist could
produce the precious metal; this idea extended to
the other precious metals, and, finally, was applied
to the animal kingdom, the belief being that the
human organization had also a special formative
force which, could it be controlled, would make
eternal existence and enduring youth possible; as a
consequence, their labors in quest of the "Philosopher's stone" and the "*elixir vitæ*" were persistently
pursued.

Some idea of the familiarity of these Arabian

alchemists with some of our most useful chemical compounds may be formed from the fact that Geber, who lived in the eighth century, and who contended that all metals were composed of mercury and sulphur combining in different proportions, is accredited with having produced nitric acid, hydrochloric acid, corrosive sublimate, and red precipitate; and we also find the early Arabian physicians with sulphuret of arsenic, sulphate of copper, sulphate of iron, and borax among their medicinal agents; besides this, when we meet with such of our drugs as alcohol, naptha, and camphor, bearing the marks of their christening, together with the many others before mentioned, we cannot but feel that but for one primordial misfortune the history of Arabic medicine would have shone with an effulgence that would have eclipsed the Alexandrian school—that misfortune being the application, in true Galenic style, of the speculative method of explaining nature, taking the Galenic errors of fact and doctrine as essential postulates, and from them producing more extended ones of the falsity that may be presented as the fruits of purely speculative and imaginative methods of ratiocination.

It would not be in place to give mention to these

and other matters transpiring in Saracenic medicine
were it not that a correct understanding of their
bearing upon our subject renders it essential; the
same may be said of those synchronal incidents of a
general historical character, some reference to which
was unavoidable if we would have a fair appreciation of those underlying causes which wrought
change at this epoch in the medical art.

The several physicians prominent in Arabic medicine, beginning with Rhazes in the east, and ending
with Avenzoar and Averroes of the Saracenic school
of Spain, will not, therefore, receive individual attention. Neither will their astrology (though enhanced by them, and employed both separately
as well as in its assumed application to the healing
art,) receive more than this passing acknowledgment.

With what has been said concerning the Arabian
age of civilization we will now turn to Europe in
tracing the devious course of medical progress. The
deserts of Tartary furnishing the barbarous Turks
to effect the conquest of the Arabian dynasty and
the consequent dispersion of science in the east,
leave only Spain under the rule of the Saracens.
Presently the Arabs, with the wealth and civiliza-

tion acquired here by them are, together with the
brilliant school of Cardova, despoiled and overthrown
by the Vandalic Christians, while the Saracens
themselves are hurled from their last land of conquest. This was the final act which contributed to
the placing of the whole Christian world under the
gloom and darkness of the most complete superstition and gothicism.

.

CHAPTER V.

The state of medicine in christendom, during the
period of Arabic prosperity, had been one of retro-
gression—as has already been observed in these
pages—the methods of medicine in treating disease
being limited to the use of relics of martyrs, ceremonial observances, supplicatory entreaties to deity,
and the like.

This condition of the art continued supreme until
about the twelfth century, although it was not
wholly relinquished then, for the darkness of apathic
ignorance which settled upon the world in the
eighth was not lifted until far into the fifteenth
century.

During this age the management of medicine rested
with the priests or their immediate subordinates.
The relics of martyrs and importunities that the good
offices of the church be extended were the sole expedients resorted to. After a time, however, and partly,
at least, through the commercial intercourse of the

(83)

Jews between the different nations, by which means some knowledge of the Arabian and Grecian works was acquired, attention began to be turned to the translation of these writings, and, later on, the wars of the Crusaders contributed to this incentive, so that at length we find the convents and ecclesiastical bodies giving their sanction, and supervising and offering facilities for instruction in "physic," and that, too, within the pale of their sacred investitures.

The unvaried sonaunce in the medical events during the dark ages is at length relieved by the work of the medical school of Salernum. This city, in the south of Italy, being in the way of the most convenient route from Europe to Palestine, was the point of departure by water to Asia for the Crusaders, and this is one reason why Salernum and its medical schools acquired prominence.

In the tenth century even the afflicted were attracted to this beautiful haven, desiring to enlist those of the medical school of Salerno in their cause. At this time only the efficacy of prayer and atonements were employed, but soon the translation of the Arabic and Grecian medical works led to the use of other modes of treatment of disease in conjunction with the above.

The merit and repute of this medical school continued to increase from this time, it being the first to extend a prescribed course of study to its students and to attest therefor by proper credentials as to the satisfactory completion of such a course. The greatest celebrity of this school was reached in the thirteenth century, when Frederick II, King of Naples and Sicily, by his systematic and substantial encouragement placed it under the most auspicious circumstances possible. Other kings were prompted to a like action, and soon the universities of Paris and Bologna excelled and obscured this the most celebrated medical school of the Dark Ages.

From this time we shall find schools of medicine in operation, with regular and prescribed courses of instruction to offer, and conferring upon those who satisfactorily completed such courses the titles of licentiate, master, and bachelor in this art. The dominant school in medicine reasserts itself in Europe at this point in the control of all matters medical, and assuming charge of the medical universities, which, under the patronage of the different sovereigns, prospered and multiplied, reiterates its allegiance to the *ex cathedra* of Aristotle and Galen by terming itself the Galenic School of Medicine.

True to the genius of their archetype these secta-tors transplanted Arabian astrology to their system of medicine, urging this absurdity as the doctrine of medicine, in the light of which medical facts should be explained and medical treatment ex-hibited.

Their theory of simple and judicial astrology claimed to be able to foresee and account for the occurrences in a human lifetime, occasioned by the influence (because of their alleged relative position at a stated time) of the sun, stars, and planets, these bodies being supposed to be possessed with the ability to issue certain influences from themselves which would effect fortunate or unfortunate results, as the case might be.

The Hippocratic axiom, that only the result of repeated actual observation was predicable, was here repudiated, and the Aristotlean and Galenic method of assuming the impossibility of error among its exponents, and of asserting the power of evolving from the subtleties of their imagination or inner consciousness a full and perfect interpretation of natural laws, was here brought into prominence, as it had been in so many former instances. The result was that this system of astrology was generally accepted, in its full bearing upon medical thought, in the different schools of medicine throughout Europe at this time.

The greatest service rendered to medicine at this period was by one Andreas Vesalius, whose researches and labors resulted in the founding of the modern science of anatomy. As a youth he displayed a great thirst and appetency for anatomical knowledge. Though, immediately preceding his time, several dissections of the human body had been made, yet these had been, to a great extent, unsatisfactory, for the Italian professors, who had undertaken these dissections, had aimed to make such investigation subservient to, and but an indorsement of, the dicta and authority of Galen.

Vesalius placed himself under the best advantages that the ablest teachers and most excellent schools could offer, but could but be disappointed with what he received. He then took most extraordinary means for following his natural inclination in his search for anatomical facts. He struggled both with bird and canine vultures for the bodies of criminals, and disinterred bodies from the cemeteries "at the risk of incurring the accusation of the capital crime of sacrilege."

He suffered even more than this, for he was condemned to death at the hands of the Spanish Inquisition, but he persevered in his seeking for anatomical knowledge. By publishing treatises, he contended and also demonstrated that Galen's opinions and teachings were false, that any one could, by actual observation, give the same evidence as he in the premises.

Because he took this course he encountered the fiercest opposition and stigma of the dominant Galenic school, and also because he " threw off this yoke of authority which had been imposed by a blind veneration of the opinion of the ancients, and ventured to conceive the possibility of error in the writings of Galen," he was upbraided and denounced as a schismatic and heretic unworthy of confidence, upon whom it was justifiable to heap the greatest obloquy.

He met the issue squarely, however, and to him more than to any other man is due the credit of having proved by his triumphs that, as regards the facts of science, the unsupported assertions of Galen, Aristotle, or any of the other ancient moguls of Dogmatism, were neither oracles nor necessarily true because of their origin.

If the war of Vesalius upon the tyranny of Galenicism settled the question of the probability of inaccuracy with the ancient lights of Dogmatism as regards their indoctrinations and methods in anatomy, it also prevailed with equal force concerning their fallibility in the other medical sciences, establishing it as clearly in physiology as in anatomy, that actual observation and demonstration were more competent than the cabal of Galen or Aristotle in verifying a fact or method before the medical world.

Contemporaneously with Vesalius there appears one Paracelsus who is also a laborer in the medical vineyard, the record of whose work bears out the claim, made for his memory, of the contribution of a most useful and advantageous life to the cause of reformation in the medical art.

Under the title of Paracelsus this man—the son of a physician, a native of Switzerland, educated by his father, and then by him placed in the several leading medical schools of his day—became a person of most pronounced talents and independence. He defied the censure and imprecations of his Galenic opponents, and in his written efforts, as well as in public, assailed the bumptious egotism of the

Galenists, and the self-satisfaction with which they viewed their own defective and erroneous methods and doctrines in medicine.

His hostility to the Galenic theory of Humoral Pathology, his antagonism to their strong evacuants, and the redundancy of drugs in the mixtures and prescriptions of the dominant school, his familiarity with and new views upon dietetics, and his introduction of the chemical substances into the materia medica would have proved either one of them to have been an efficient reason for the visitation upon him of the combined gall, anathema, and malignity of the locum tenens of Galenicism.

The righteousness of the cause of this doughty warrior gave him a like advantage in the province of medicinal agents and their application to disease, over his opponents, that Vesalius obtained in certain directions, and his blows against Galenicism were no less effective than those of the rare anatomist of the sixteenth century.

Later, when the dominant school renounces its alchemic and theosophic tenets, we find it, in its assaults against the influence and work of Paracelsus, pronouncing upon him the epithet of astrologist and alchemist, although his writings show him to have antagonized the claims of astrology, and to have neither sympathized with the Alchemists, nor to have been recognized as among their number by the Alchemists themselves. The fact that he wrote and successfully contended for the recognition and use of chemicals is the only conceivable apology for the appearance of the latter charge.

For many years allusion to Paracelsus or his labors in medicine created very much the same effect upon the adherent to Galenicism as do the bright colored garments of the *chulos* upon their victim in the *fiesta de toros*.

This school has raised, argued, and decided, much to its satisfaction, the question of the erratic character and immoral and dissolute habits of Paracelsus, and assert that he was possessed of most consummate self-conceit, and cite as evidence in this that, while he was professor of medicine and surgery at Bale, he publicly burned the writings of Galen, and was compelled to flee the country therefor shortly afterward.

Their claim that he pretended to have discovered the *elixir vitæ* is dwelt upon in conjunction with the asseveration of his death in a hospital at the age of forty-eight from indulgences in a period of drunk-

enness and debauchery. However, these attacks
upon his personal character do not affect the quality
or value of his work in its relation to medical reformation; neither does the fact that he occupied an
onerous position before the world, when to take his
independent stand was to suffer persecution and
misrepresentation, appear any the less great. Moreover the attempt to show that he was anything else
than a great man—when great men were few—
proves to be but abortive and unsuccessful.

The empiricism of Paracelsus, though so often
antagonized by the Galenists, was nevertheless
founded upon principles of a more enduring character than the hypothetical physiological elements
of Galen, for careful observation was made his basis
as this quotation from him will evidence. "There
are two kinds of ways and paths, or two methods and
fashions, by which to arrive at a knowledge of the
arts. The one teaches and conducts to the truth;
the other to illusion. The erring and vagabond
discourses on the understanding and reason are the
causes of error, which occurs when they are permitted to guide us. Experience and that which is
found familiar and in accordance with nature, and
which produces like actions, is the cause of truth and
certainty."

The bias and tendencies of the dominant school were consistently pushed to their ultimates, and we therefore find the adherents of this school assuming for themselves, as Galenists, most extraordinary endowments, for from proclaiming the ability to produce imaginative hypotheses concerning which they claimed to present the true and only explication, we now find that they assert their derivation of knowledge of mooted matters in medicine to be from the Deity—to be a matter of inspiration. These assumptions were those of the Galenic school at this time, and were by its representatives very generally accepted and assumed.

However these alleged powers were not to any extent applied to anatomy or the allied sciences, for this field was occupied at this time by an indefatigable force of followers of Vesalius, the results of whose efforts had placed these branches beyond the limits of Dogmatic control, only leaving the Galenists the intricacies and marvels of disease phenomena, and the methods and manner of the action of curative means to hypothecate upon.

Naturally the course taken by the Galenists, as above described, led to the promulgation of the doctrine that disease was caused and controlled by

demons; and contrary to common sense as this preposterous doctrine strikes us to be, yet it was the
doctrine which spread and for years held the medical schools of Europe.

Painful as this condition of things was concerning
the nature and treatment of disease, we shall turn
with a feeling of relief to ascertain of the state
of anatomy and physiology at this time, after
Vesalius, Fallopius, Eustachius, and others made
many discoveries in anatomy, contributing much to
its advancement. We also find Fabricius describing
the values of the veins, and William Harvey making
known the mechanism of the greater and lesser
circulation. These with the deportment of the blood
corpuscles in the capillary circulation, which, by the
aid of the microscope, was demonstrated by Malpighi,
exampled an aggregate of progress in marked contrast to those branches of medicine still under
Galenic method of ratiocination, though the other
sciences, as chemistry and astronomy, are found to
have kept pace with anatomy in their measure of
general advancement.

It would seem, as if when this amalgamated school
of fallacy, superstition, and malignancy, had, from
the force of external influence, abandoned its last

absurdity—for through the printing press and the
Reformation congeric forces had arisen—in the
memory of its past blunders and errors, both of
omission and commission, it would for the future
engage a more tolerant bearing toward those who
declined to accept the infallibility of its imaginative
powers of dicta.

Much as this was to be hoped for, we do not, however, find it to be the case; for if, as we have seen it
down from the ancient days of Aristotle and Galen,
on through the dark ages to this seventeenth century
the inborn bent of this school has been to procreate
a chimera which shall engross their obeisance and
fealty for a brief time until its instability explodes
it, leaving its worshippers to bring forward newer
and even greater imaginative follies, only to receive
in turn the same experience at the hands of Dogmatic medicine—*uno avulso non deficit alter.**

The conceit of this school will also be found to
continue on, even down to the present moment,
always occupying the same position of distinction
and honor; a conceit which maintains that it shall
reside supreme within the faculty of Dogmatic medi-

---

* When one is torn away another succeeds.

cine, to be known as its own prerogative and endowed efficient genius; a conceit which declares and affirms from the fullness of this assumed ability and autocracy what shall be the views and interpretation of such phenomena as have bearing upon the medical art, and what shall be the accepted theories and predictions concerning them.

This is the worst attribute of the dominant school in medicine, and also its most prominent one. A due appreciation of this trait will explain the source of the misdirection given to thought in its attempts to develop the medical art. We shall give but little attention to those fantasies, the controling and directing theories of the dominant school, that appear in the later centuries, believing that to make more reference to them than the claims of our subject demand will prove both tiresome and unnecessary.

We now turn our attention to the career of an English physician whose talents and distinction in his profession, together with his efforts to raise the status of medical science, are deserving of a prominent place in history.

Thomas Sydenham was born in 1624, at Dorsetshire, England, he was educated at Oxford. His degree in medicine was the only degree he ever received. He went to Montpelier, in France, where he followed the methods and inculcations of the professors of the medical school in that place.

In 1660 he settled in London, to undertake the practice of his profession. His brother physicians did not regard him kindly, and while settled here, his reception by them was always marked by this feeling on their part. The College of Physicians declined to receive him as a fellow, making him a licentiate only. However, his marked abilities and very extensive practice, soon placed him in a most prominent position before the profession. He has been styled the English Hippocrates, and he shows us more points of likeness to the " Father of Medicine " than any one that has gone before him. His writings give us many theoretical concepts of the nature and causes of disease and of other matters.

Both the ancient and English Hippocrates being human, the influences of their time and surroundings contaminated them, though they each produced theories which unfortunately appear not to their credit, yet these theories and their acceptance by the profession were not their absorbing aim but matters of minor importance. .

7

Their great work and aim was the careful interrogation of nature with a view of studying the
unfolding of the manifold forms and features of
different diseases, that their true nature might be
understood. Thus, Sydenham's aim was, by devoting himself to patient and careful research, and
trusting that the result might prove of some additional help to suffering humanity, to allow the great
theoretical follies and all that pertained to them to
be monopolized by the College of Physicians. This
course of his carried him on to another work which
he now instituted, and in so doing opened the way
for one to follow who should bequeath an im-
measureable benefit to mankind.

The particular attention paid by Sydenham to the
study of symptoms and their significance—together
with his careful following of the effects of medicine
upon these symptoms—were of great value to the
medical art, enabling him to show the great efficacy
of cinchona in fever and of cooling remedies in
small-pox. The superiority of his new method of
treatment of this disease over the old methods saved
many thousands of lives in his day.

After his death, his fame and work were exalted,
and himself held up as the typical representative of

Dogmatism. In fact, however, these innovations and valuable researches of Sydenham in his labors for the advancement and promotion of medicine were at the time adversely received by Dogmatism, the antagonism and execration of this school being visited upon him, though afterwards these researches reverted to the honor of Dogmatism, while on the other hand the contagion of its tendencies were unfortunately to some extent communicated to Sydenham himself as shown by his own writings. In contradistinction, however, to his school, he declined to urge his views upon the profession, showing instead a remarkable tolerant spirit. In this century the old and absurd theories of the Chymical sect were revived, and, with some amplification, became the accepted system of Galenicism, though in the last part of the seventeenth century the Iatro-mathematical hypothesis succeeded this.

George Ernst Stahl, born in Germany, in this century, became known to the medical world in the beginning of the eighteenth. Of powerful and brilliant mental mould, from a familiarity with the later theories, and a study of the writings of Von Helmont, as well as those of Becher, the physicist, Stahl, from his own reflections, constructed a theory concerning the nature and cause of the phenomena of life. He held that every change, and, therefore every disease was caused by an *anima* or spiritual principle which dwelt in the body, producing its development and controlling its powers and functions.

For a long time this doctrine was very generally accepted by the dominant school, its influence and logical sequences producing another great change upon the methods of medical treatment. Stahl distinguished himself in other directions, giving fresh impetus to chemistry by furnishing the phlogiston theory. He died in 1734.

Contemporaneously with Stahl, Herman Bœrhaave occupied a place in the medical profession which, from its eminence and from his own personal popularity, went to make his a most influential position. He has sometimes, because of his combined incongruities and brilliancy of parts, been termed an eclectic. He died in 1738.

In the year 1735, John Brown, an afterward noted physician, was born of obscure Scottish parents. He was a pupil of Cullen and the originator of a system of medicine which, in its turn, received prominence at the hands of the dominant school.

This system taught that life is only sustained by incitation, " It is only the result of the action of incitants upon the incitability of organs." It also held that the only pathological condition was one of sthenia, or asthenia. In the main, the treatment used in this Brunoian system was, on the one hand, stimulants, and, on the other, bleeding, purgatives, and opium.

We now will engage in an unusually lengthy consideration of the salient points in the history, as well as the work undertaken and accomplished by one of the most extraordinary men that has adorned the realm of medicine. The call for this rather lengthy review will be apparent before we have finished, and it will suffice if we here state that it concerns one who founded a rival school of medicine, which, to this day, occupies the field with Dogmatism, contesting and contending against this school at nearly every step, and, in this way, to an extent that no other adversary of Dogmatism has done before her, coerced and constrained the dominant school to vacate its injurious tenets and for humanity's sake, pursue new and more desirable methods.

Samuel Hahnemann was born in Saxony, Germany, in 1755. As a boy, Hahnemann displayed a great desire and an equal aptitude for study, but his father being a porcelain painter, and in poor circumstances, decidedly discouraged his longing for the higher branches of knowledge. However, we learn that he disobeyed his parents to that extent, that he designs and produces a lamp which shall assist him to gratify his craving for knowledge with safety while the others of the family are wrapt in sleep. His schoolmaster, having a great admiration for his unusual abilities, without remuneration, took the young prodigy and gave him the desired instruction, and only concluded when his pupil reached his twentieth year. "On leaving school it was the custom to write an essay on some subject, and Hahneman selected the somewhat unusual one of " the wonderful structure of the human hand." Who would not like to see how the boy Hahnemann treated this subject, his selection of which shows a strong bias towards natural science."* From the hands of his benevolent schoolmaster, he was able to enter the University of Leipsic, where, in order that he might procure the necessaries of life while pursuing the

---

* Dudgeon.

curriculum of study, he took private pupils to instruct in French, English, and German.

He went to Vienna subsequently that he might devote himself to medicine and follow in the hospitals the practical as well as the theoretical teachings of his chosen profession; his limited means failed him, however, before completing his courses of study, and therefore he was only too willing to accept the position offered him of family physician and librarian to the Governor of Transylvania. He had been fortunate enough at Vienna to secure the friendship of the great Dr. Von Quain, who, very probably, was the means of his receiving at this time this, to him, most fortunate offer.

After remaining here two years he repaired to the University of Erlangen, where he graduated in medicine in 1779. From here he returned to his native land, Saxony, where he settled in a small country town to practice his profession, and shortly after accepted the position of district physician. Three years from this time he relinquished his practice in Gommern, to go to Dresden, where he served for a year as physician to the hospital.

Contemplating now, in the light of his last eight years experience in medicine, the darkness and

uncertainty of medical methods, and reflecting concerning those fallacious and delusive means which
were the regularly enjoined practice of orthodox
medicine, and which abounded in those unfortunate
results, the natural outcome of the strict following
of these prescribed methods, witnessed by Hahnemann himself, both in his own practice as well as
in the practice of his profession in the hospitals and
also in that of his colleagues, he found himself
acquiring a strong abhorrence and disgust of the
practice of medicine as understood and followed at
this time.

To quote his own words in reference to the result
of his reflections and cogitations: " It was painful
to me to grope in the dark, guided only by our
books on the treatment of the sick—to prescribe,
according to this or that (*fanciful*) view of the
nature of diseases, substances that owed only to
mere opinion their place in the materia medica ; I
had conscientious scruples about treating unknown
morbid states in my suffering fellow-creatures with
these unknown medicines, which, being powerful
substances, may, if they were not *exactly* suitable,
(and how could the physician know whether they
were suitable or not, seeing that their peculiar

special actions were not yet elucidated?) easily change life into death, or produce new affections or chronic ailments, which are often much more difficult to remove than the original disease. To become in this way a murderer or aggravator of the sufferings of my brethren of mankind, was to me a fearful thought—so fearful and distressing was it, that shortly after my marriage I abandoned practice, and scarcely treated any one for fear of doing him harm, and, as you know, occupied myself with chemistry and literary labors."

Having now retired from active life, and residing in the adjacent village of Lockowitz, near Dresden, we may believe that, in the privacy of solitude, his deliberations and meditations were productive of results which, in future years, stood him in good stead in the elaboration of his original and useful discoveries.

During this hermitage, however, he devoted much patient work and study to the science of chemistry, and in the course of the next few years many valuable papers were given to the world from his pen, which establish at once his talents and great sagacity as well as his profound and thorough knowledge of chemical facts and processes.

Of these valuable papers of his upon chemical
subjects, which may be counted by the dozen, those
upon the deportment of metallic substances and
upon the nature and art of testing wines attracted
noticeable attention for many years, while his treatment of the subject of "*Poisoning by arsenic, the
remedies for it, and its medico-legal investigation*," was
a model of acumen and power, and has, until a
recent day, been an authority often quoted.

Beside these literary efforts, previous to the year
1790, he made translations from French, English,
Latin, Greek, Syriac, and Hebrew, upon divers subjects, such as Medicine, Physiology, Chemistry, the
Arts, etc., these manifold efforts stamping him as a
man of breadth and erudition. We may believe
that during this period of his life Hahnemann,
though not engaging in the practice of medicine,
often felt a yearning for it and its associations, and
at such times, or when illness appeared within his
own household, the question would return to him
with added force—why is not the noblest art of man
reclaimed from its gropings in darkness? Why has
she not a sure and guiding principle, which, as a
trusty star in the east, shall pilot her on the royal
highway to certainty and success in the cause of
suffering humanity?

Filled with these longings and freed from prejudice, his powerful mental faculties were in an acute and receptive state. In 1790, when engaged upon a translation of Dr. Cullen's Materia Medica, while engrossed with the drug cinchona, he was struck with the fever condition which Cullen stated was produced by it. Having had intermittent fever himself some years previously, and seeing a similarity between the cinchona symptoms of Cullen's, and the symptoms of his own sickness cured by this medicine, he set about experimenting relative to its effects upon the healthy human system.

This fact was to him " What the falling apple was to Newton and the swinging lamp in the Baptistery at Pisa to Galileo." The results of his observation of the action of Peruvian Bark upon the healthy organism, was the discovery of the principle of drug action, and, therefore, a scientific basis for their medical employment.

For several years Hahnemann made no effort to make known his newly acquired knowledge. Instead of this, when in 1792, he was offered the trust and management of the Insane Asylum in Georgethal, which was accompanied with a fulsome salary, he gladly accepted, seeing opportunity here to pursue and verify his new discoveries.

We see this sagacious man instituting here radical
improvements outside of the field of medication by
originating and employing the moral treatment of
the insane, he says: "I never allow any insane person to be punished by blows or other painful cor-
poral inflictions, since there can be no punishment
where there is no sense of responsibility; and since
such patients cannot be improved but must be
rendered worse by such rough treatment." As it
was in the latter part of 1792, when Pinel, who
claims to be the first, tried this method with the
insane we may safely assume that to Hahnemann
belongs equally the honor of bringing forward what
has since proven to be a great innovation in the
treatment of this unfortunate class.

The complete and happy recovery of the Hano-
verian Minister brought the name of Hahnemann
and the institution prominently before the public.
In 1796 there appeared from the pen of Hahnemann, in the then leading medical journal in
Germany, (Hufelands,) an article under the title,
"On a new Principle for ascertaining the Remedial
Powers of Medicinal Substances." In this paper he
emphatically states that chronic diseases are better
cured by those medicines which produce a similar

action upon the healthy human organism, and demands that remedies be proved upon the healthy, and also says: "Every efficacious remedy produces in the human body a peculiar species of disease; and the more powerful the remedy the more peculiarly distinguished and severe is the disease produced."

At intervals through this decade appeared also from his pen "Are the Impediments to Certainty and Simplicity in Practical Medicine insurmountable?" (1797.) "A case of Colicodynia quickly cured." "Antidotes for some Heroic Medicinal Substances." (1798.) "Some kinds of Continuous and Remittent fevers." "Several Periodical and Hebdomadal diseases." "Observations on the Three Current Modes of Treatment." (1801.)

These convictions aroused at once most unfriendly criticism, displaying feelings of animosity towards their author, and continuing uninterruptedly. Some years previous to 1796, Hahnemann, in a preface written by himself to an English work which he had translated says, that the "original informs me, even in London, medical frankness requires the ægis of anonymousness in order to escape being chid."

However, the numerous attacks upon his opinions,

as well as the reproach and abuse which was
launched against him, directed attention to his deductions, with the result that men of ability and
high standing were attached to them, and espoused
them, and added their efforts to his to defend his
position before the medical world.

Previous to the year 1801 Hahnemann had advocated *reform only* in medicine, attempting, by investigation and experiment, to lead the dominant school
to renounce their present fallacies, as taught in their
then accepted theories of Brunonism, and to investigate the results of his inquiries in the action of
drugs, as presented in his many treatises upon this
subject, and to pass upon these fairly.

Remembering the emphatic and denunciatory
judgments of some of the most prominent physicians in the dominant school regarding the competency of the Dogmatic methods, and, finding his
experience to more than corroborate these opinions,
he probably thought this call for reform not unrea-
sonable.* He, therefore, claimed liberty of medical
opinion and action, and concluded his essay on
"View of Professional Liberality at the commencement of the Nineteenth Century," (1801,) by adjuring

---

* See foot note, page 44.

his colleagues in these words: "Physicians of Germany *be brothers*, be just." In pursuing his observations and study of drugs, in accordance with his views as published in 1796, that "the true remedial powers of a medicine for chronic disease" could only be known by ascertaining its action upon the healthy body, and that, to overcome a curable disease in a safe and speedy manner, the drug must be used that produces the nearest "similar morbid condition," he met with a measure of success which he had hardly dared to hope for, and, very naturally, ceased the making of prescriptions containing many drugs, but used that drug, in accordance with his new principle, which had produced the nearest similar action.

Success in his new treatment naturally gave him a rapid and increasing reputation, and having reason to believe that the apothecaries did not furnish as pure drugs to his patients as they ought, wishing to feel sure of the cause of failures in his practice, he began to prepare his own drugs.

The physicians of Königslutter, where he now had a large and growing practice, who were doubtless jealous and unwilling to be placed where the skill of Hahnemann could be a criterion by which their own success might be judged, spurred on the apothecaries to harass him by claiming that he was transgressing the law by the compounding and selling of prescriptions.

He answered that his prescriptions were uncompounded drugs, that he gave but one medicine at a time, and that the spirit of the law intended to discriminate against the prosecution of the drug business by the ignorant, and that the medical man, in cases of emergency or for other reasons, should enjoy that privilege. However, this was decided against him, and being too honorable to prescribe in secret or to down his conscience and place his patients in the hands of the apothecary, he resolved to leave his large and increasing clientage and many dear friends, to seek more congenial surroundings for the prosecution of his life work.

His new methods had now developed so far that he was able, in a very short time, to establish a great reputation by the comparison of his results at the sick bed with those of the other physicians. He was dreaded alike by the physicians and apothecaries, with the result that we find poor Hahnemann driven from pillar to post.

He left Königslutter in 1799, and within the next

two years he abandoned successively, five or six of the different towns in Germany where he had attempted to establish himself. Such persecution as this added to the cares and responsibilities of a large family, together with the personal opprobrium with which his traducers loaded him, must have exasperated and soured his naturally benevolent and sweet disposition.

From the time of his being compelled to leave Eulenberg, in Saxony, on account of the persecution of the superintending physician there, we may believe that he became more and more discouraged at the possibility of not achieving what he had been so many years striving for—a reform of Dogmatic medicine.

Therefore, when in the continued experience and observation of the true action of medicine, he sees enhanced results from the most ultimate division and dilution of medicines, and announces this new discovery with the result that more ridicule and anathema than ever is heaped upon his devoted head, we may understand why, after, in 1801, having, in a most convincing and effectual essay on Brunonism, clearly exposed the absurdities, inconsistencies, and dangerous methods of this system, he,

S

in 1805, wrote two powerful papers on "Esculapius in the balance" and "The Medicine of Experience" which foreshadow those intentions, now fast assuming definite shape, of severing his connection with the Dogmatic school, that in freedom from restraint, he may perfect and develop his new methods into a new system of medicine.

Previous to 1808 Hahnemann spoke of medicines exhibiting "specific action," and he employed this term in connection with his use of medicine in the light of their "similar" properties. After the year 1808, however, he gives us the word homœopathic, as applied both to the action of remedies and to his new principle or law, terming it the "Homœopathic Law of Cure." In these essays during this year and in many other papers, which in the next two years appeared from his prolific pen, he is constantly bringing to view the development of his new system, until in the year 1810 he fully, and after the aphoristic method, sets forth in an elaborate treatise ("The Organon of the Rational Art of Healing,") his new system, its preëminent excellence as compared to the old, and its direct antagonistic position to the instincts and genius of the dominant school.

In doing this he gives these definitions of the

systems of medicine. He defines the "homœopathic" action of a drug to be an action that is as near as possible similar to any particular disease, and the term "allopathic" action to include an effect which is not only opposite to that of the disease but including everything essentially of another character, even if not strictly opposite. Hence he who accepts and applies this homœopathic action of drugs in disease, and therefore indorses this essential law of cure, "*Similia Similibus Curantur*," becomes identified with the Homœopathic school of practice, while he who endeavors to combat disease by employing agents or medicines that produce another or opposite disease and cycle of symptoms is in the recognized practice of the dominant school which, from the words ἄλλος, another, and πάθος, affection, or its English *allopathy* is called by Hahnemann the Allopathic School in contradistinction to the Homœopathic School of Medicine.

"Without ever regarding that which is really diseased in the body it (allopathy) attacks those parts which are sound, in order to draw off the malady from another quarter and direct it towards the latter."* "The pernicious results of such a prac-

---

* Organon.

tice, at variance with nature and experience, may
be easily imagined."*   If, however, now, from the
malignant personal, as well as other unfair attacks,
he is exasperated and irritated beyond measure, and
exhibits an inclination to answer the Dogmatic
assertions concerning medical matters in a like
Dogmatic spirit, these things, with other lapses from
his true genius, must be taken as human, and considered in connection with the attending circumstances, for these lapses are but ephemeral in practice and but a shibboleth to the future adherents of
this school of medicine.   .

In the Organon his thorough and studied exposition of the homœopathic method is clear and well
defined.   Claiming that it is the duty of the physician to "search after that which is to be cured in
the disease," and so acquaint himself with all the
presenting symptoms, he declares, first, that, "in the
cure of disease, it is necessary to regard the funda-
mental cause and other circumstances," and that
"the three necessary points in reality are, to ascertain the malady ; second, the action of medicines ;
and, third, their appropriate application."   Besides
this, he lays stress upon a proper regard for diet and
surroundings, and also submits, "In disease the

---

* Organon.

vital power is primarily disturbed, and expresses its internal changes (sufferings) by abnormal alterations in the sensations and actions of the system."

These quotations are from the Organon which further declares that, by the results of the proving of the action of drugs upon the healthy, there are shown such effects upon the "vital powers" as to beget diseases, or cycles of symptoms which closely simulate certain diseases, and that the exhibition of this remedy, when it is the closest similar symptom producing medicine, to a given disease, will demonstrate an unequalled effect for cure. To express this action or power of drugs the law *similia similibus curantur* was brought forth, this being now the essential law and corner-stone of Homœopathy.

To administer several drugs together, each of which had received separate study by proving, would not be in conformity with this principle, for many of these medicines are direct antidotes to each other, and also indirectly so upon certain of their single symptoms, and, therefore, the results would be confusing, and such practice a blind one. Because of these considerations, and where homœopathy is applicable, only that drug should be applied against disease that we know to be its nearest similar ; this led to the assertion of the axiom that only one drug should be administered at a time, and not a mixture of many, until reason for believing that another medicine was better indicated, when it should be substituted.

In the continued observation of this law an additional feature was also developed. In the employment of the crude drugs, after their homœopathic indications, a varying and uncertain degree of impression upon the patient was produced, particular attention was given to this fact and led to much thought and study concerning the most favorable method of preparing medicines that a definite and reliable action might be secured.

In a paper published in 1801, "On the Power of Small Doses of Medicine," etc., Hahnemann manifests already positive ideas upon this subject, he says of pills, bolus, and the like, as then prepared for administration: "It (the pill) slides almost completely undissolved over the surface of the intestinal canal, invested with a layer of mucus, until it (in this manner itself covered with mucus) completely buried in excrement, is speedily expelled in the natural manner;" we may understand why this eminent chemist, pharmaceutist, and physician devoted himself to finding an improved method of preparing drugs.

Finally, he announces that the particles of a solid medicine should be separated from themselves by being thoroughly mixed by grinding and incorporation with a substance which is both inert upon the human system and easily assimilated by it. Sugar, which is contained in the milk of the goat, being obtained from it, displays these qualities, and is, with these solid drugs, by the rules and processes of Hahnemann's method of trituration, the vehicle employed. With liquid drugs, such as tinctures of plants the method of dilution by succussion, etc., which Hahnemann directs, is employed.

Hahnemann claimed for these simple methods a presenting of many hundred times more surface of a given quantity of a drug to the surface of the assimilative parts, and also by this minute subdivision of the particles of a drug and their association with this excellent vehicle (sugar of milk) to render it more thoroughly assimilative by all the fluid and solid parts of the body and therefore more efficient in impressing the system.

Hahnemann and his followers were well repaid for their researches and patience with this question, by obtaining more reliable and vastly more brilliant results from the use of such methods in preparing their medicines. It is well known that many people display idiosyncracies or peculiar sensitiveness toward the action of different drugs, one person with one, and another with a different drug. Complaints from such persons of suffering from the effects of a medicine were, at his time, and are to-day, a common experience with physicians. For this reason, and also because he believed that medicines, by their crude strength, often aggravate for a time, at least, the disease under treatment, we find Hahnemann continuing to attenuate still higher, by his methods just mentioned, his medicines, and with the consequence that better and happier results in treatment were obtained than heretofore. Seeing in these results substantiating evidence of the value of the similar action of drugs in overcoming disease, and believing that such a preparation of medicine would only impress the system by acting upon the vital power, and through its influence there produce this train of symptoms, which were *similar* to those produced by the disease, (for the physiological or drug effect produced by the disease could not have here caused this curative

influence, for the dose, by the method of administration, contradicted this,) he heralds this additional
feature as a most effective reinforcement of the
Homœopathic method of cure.

Briefly, then, his Organon presented the "great law
of cure," *similia similibus curantur*, as the important
essential and necessary principle of Homœopathy,
and the sequential features, the natural outcome
of this, appear : First, the proving of drugs upon
the healthy. Second, the administration of a drug
singly that we may act intelligently and profitably.
Third, the preparation of remedies according to the
homœopathic methods of trituration and dilution,
this being a substantial and most needed help to the
essential features of homœopathy in their application
to disease.

In the next year, 1811, appeared the first volume
of Hahnemann's Materia Medica Pura, in which
were collected those provings, which he, in late
years, had been so diligent in procuring. This
book was welcomed and studied with much eager-
ness.

Desiring now to place himself in a position where
he could present his new methods of medical practice to physicians and medical students for their

consideration and investigation, in the words of Dr.
Dudgeon, " he resolved to give a course of lectures
upon the system to those medical men and students
who wished to be instructed in it. In order to be
allowed to do this, however, he had to pay a certain
sum of money, and defend a thesis before the Faculty
of Medicine. To this regulation we are indebted
for that able essay, *De Helleborismo veterum*, which
no one can read without confessing that Hahnemann treats the subject in a masterly way, and displays an amount of acquaintance with the writings
of the Greek, Latin, Arabic, and other physicians,
from Hippocrates down to his own time, that is
possessed by few, and a power of philological criticism that has been rarely equalled. This thesis he
defended on the 26th of June, 1812, and it drew from
his adversaries an unwilling acknowledgment of
his learning and genius, and from the impartial and
worthy dean of the faculty a strong expression of
admiration. When a candidate defends his thesis,
he has what are called opponents among the examiners, who dispute the various opinions broached in
the thesis; but the most of Hahnemann's opponents
were schooled into such an amiable state of mind
by this display of learning, that they hastened to

confess they were entirely of his way of thinking, while a few, who wished to say something for form's sake, merely expressed their dissent from some of Hahnemann's philological views. This trial, which his enemies had fain hoped would end in an exposure of the shallow charlatan, triumphantly proved the superiority of Hahnemann over his opponents, even in their own territory, and was a brilliant inauguration of the lectures which he forthwith commenced to deliver to a circle of admiring students and grey-headed old doctors, whom the fame of his doctrines and his learning attracted round him."

From the year 1808, on through the intervening years to 1845, the year of his death, we find works, monographs and papers, appearing in the defence or elaboration of Homœopathy. Now, (1810,) with the appearance of many able physicians, to be followers of his, and zealous workers in the cause of Homœopathy, this new school of medicine obtains a sphere and field of usefulness broader and more extended than could be contemplated while unfolding a biographical synopsis of Hahnemann's life. Therefore, after we have considered those indictments and criminations which are inimical to

Hahnemann especially, we will turn to view the aspects of Homœopathy, its relations to its assumed object and to the dominant school in medicine, and to attend upon these with the other occurrences in the medical art of sufficient import to repay us for such study, and we shall find that, in pursuing this course, the subsequent history of Hahnemann and his achievements in the cause of homœopathy will receive as prominent notice as a work of this character will permit. We have spoken of the persecutions and detractions which followed him from the middle of the last decade of the eighteenth century and which rendered his development of his new methods so difficult, and now let us inquire into the nature of the objections made to the course which he pursued, and of the personal charges made against him.

That of being a heretic was among the earliest and probably the most frequent of the charges against him. He was guilty (and his writings proved it) of announcing opinions which did not accord with those of the established and dominant school in medicine. He had dared to evince dissatisfaction with the accepted methods of Dogmatism, had objected to them, and had endeavored to substitute other and better methods for them. These doings and expressions stamped him as a *heretic.* In looking at this indictment calmly we find that if Hahnemann spoke of the success and practices of the medical methods of Dogmatism disparagingly, and asserted dissatisfaction with them, he was but indorsing—though in no more emphatic manner— the opinions of those among the most prominent of his predecessors and contemporaries, illustrations of which we have already read in these pages.

One from the pen of him who for centuries was the great figure head of Dogmatism, and who, in exhibiting his egotistic pretensions to surpassing knowledge, says, of his fellow physicians, that they are "ignorant or punctilious dialecticians, whose discussions are frequently repugnant to the most common sense," while Hahnemann's contemporary, Bichat, tells us, "It is not a science for a methodical mind. It is a shapeless assemblage of inaccurate ideas, of observations often puerile, of deceptive remedies and formulæ as fantastically conceived as they are tediously arranged."

Hahnemann never said anything more condemnatory of the methods and skill of Dogmatic medicine than this, and we shall find, presently, that,

since this time down to the present, others have
been (mayhap when under exasperated circumstances, from the failures of the orthodox treatment,
and so betrayed into a spasm of unreserved frankness) as emphatically clear and equally as denunciatory in their conclusions concerning the merits of
Dogmatic medicine.

We need not say that none of these physicians
have been placed beyond the pale of the Dogmatic school, and pursued and maligned as
heretics because of this, so let us seek farther for
a reason for the persecution of Hahnemann. A
minor charge, not urged with the same frequency
and persistency as the last, was, that he sold secret
medicines. He did claim that he had discovered a
prophylactic and wonderful remedy for scarlet fever,
and would furnish it to physicians for trial, but
would not tell its name; but, when assailed for having a secret remedy, he immediately divulged its
name, and said that the reason why he had withheld
it was because he knew it would not receive the impartial attention that it otherwise would.

At another time he believing (by a cursory examination maybe) that he had produced a new
chemical and medicine, sold some to different per-

sons, but afterwards, finding it to be borax, made
known his mistake and promptly refunded the
money to the purchasers. We can readily understand the statement of his suffering from poverty,
when persecuted and driven from one place to
another, and who would wonder, if at such times
when thus reduced, he furnished remedies when applied to for treatment, and then exhibited some retali-
ative spirit by refusing the apothecaries and physicians a knowledge of his methods of cure? Again,
the public were repeatedly warned not to be defiled
by his touch, for he was a *quack*. This term is properly applied to an *ignorant practitioner* in medicine, if
he be honest or deceitful in his methods it matters
not, but a lack of proper previous preparation and
qualification renders him properly speaking a *quack*.
What a misnomer as applied to Hahnemann, one
of such unusual acquirements and erudition, as his
translations no less than the evidence of those most
prominent in his day in chemical science attest.

Professor Christison, referring to, quoting and
accepting Hahnemann's treatise upon arsenic, thus
making Hahnemann's opinions his own, is unstinted
in his praise of Hahnemann's chemical profundity,
as also is Professor Berzelius, who says: "That man

would have been a great chemist had he not been a great quack." Though Berzelius is competent to judge of Hahnemann's accomplishments in the natural sciences, yet we may reserve to others the judgment upon his medical qualifications. Concerning these, we may say he pursued the regular courses in Leipsic, Vienna, and Erlangen, and at the medical school of the latter city he regularly graduated in medicine, and as this school was endorsed and duly recognized by the Dogmatic faculty, we therefore see that this charge is not well founded.

If it was intended instead to attach the allied epithet of "*charlatan*," and one which we also find frequenlty applied to Hahnemann, we shall find this equally inapplicable. A charlatan, in fact, is one who applies trickery, deceit, and vain pretensions to the practice of medicine—this definition being one which, we have no doubt, will not be objected to, for it is the generally recognized meaning of the term—let us see if it is applicable to the sage of Homœopathy.

Those who have taken pains, from an interest in the matter, to inquire into the natural aptitudes and moral qualities of Hahnemann all agree that the first impression one gets concerning them is of his remarkable sweetness of disposition, this being always spoken of as his most prominent trait, and so through his youth we so often hear of his amiability and quiet mildness of manner. As we do not hear of his great brilliancy of parts, but always some reference to his sweetness of mind, together with his natural aptitude and quiet enjoyment of study, we may take these natural gifts into consideration in inquiring as to his proclivities as a man.

In every case the testimony of his biographers recount concerning his "acknowledged benevolence" and "genuine philanthropy."* Therefore, when we find that the only specification brought in support of this charge is that he insisted upon pretending to the public that his new methods were wonderfully successful, thereby drawing largely from the other physicians' practice, and because he continued to insist upon this in spite of the declaration of the Dogmatic faculty, that his methods were false and even worse; when we see that it is only upon a falling back on the infallibility of the Dogmatic assumption of the dominant school that these persecutors of Hahnemann rely for establishing their

* Marcy.

9

charge of "*charlatan*," we at once recognize that it is only after the same pattern of assault as that which this same school had, as we have seen, previously directed upon such men as Vesalius.

Moreover, every abuse of this kind was heaped, most mercilessly, upon Harvey, because of his promulgation of the real use of the blood vessels, and of the true method of the circulation of the blood, thereby offending the Dogmatic faculty by presuming to overturn their concepts and assumptions, just as Vesalius before him had done upon anatomical questions. However, we find this difference, and this only, between these great reformers and Hahnemann, that, after a time, the versatile Dogmatists retrimmed their sails and shaped the bearing of their ship of medical knowledge into a new and different course of hypothesis.

In other words, they were willing to reconsider the strictures which they had placed upon these discoverers and reformers in medicine, and would show their broad spirit of forgiveness by receiving them back with open arms into the faith, and also renounce the old and prevailing dogmas of the school and accept the new and diametrically opposite teachings of these discoverers.

This was not the case with Hahnemann's discoveries, and we shall readily see ample reasons why, to this day, he and his methods are yet antagonized, when we consider that the acceptance of these new methods would, necessarily, utterly demolish (or, at least, the supporters of the dominant school appear to think so) those interests, of a financial character, which have been, since Hahnemann's time, intimately attached to, as, in fact, they were previously created by, Dogmatism.

Another charge brought against Hahnemann, and confidently, though in the memory of previous discomfiture, is that of being a "*schismatic.*" He certainly did promote and was mainly instrumental in effecting a separation of many adherents of Dogmatism from this school of medicine to form a new school and in antagonism to it. His were not the methods of a superficial and brilliant man who would throw off the scheme of a new faction in medicine for the sake of the fame without demonstrating the need of it, or subserving the general good by so doing; to the contrary, the history of his struggles with his conscience, and then his endeavor to reform medicine, the evidence of the need of which reform was abundant as the most distinguished physicians of Dogmatism had affirmed, is a matter of record.

This record also shows that after years of patient and laborious investigation, after gradually and surely unfolding the better and far more successful methods of practice, submitting during these years to the most malicious and persistent attacks, and becoming convinced that his efforts for reform in the dominant school were and would be unavailing, (that these new principles of treatment might live and be tried upon their merits and thus serve humanity,) he severed his further connection with Dogmatism, and as we have seen, after announcing the homœopathic law, established the new school in medicine.

There are other points upon which Hahnemann was assailed, but we shall find these to have an equal or greater bearing upon homœopathy as well, and, therefore, we will now turn to the consideration of the early status and future progress of this school in medicine.

CHAPTER VI.

The Homœopathic School now became an entity indeed, for Hahnemann attracted to his side enthusiastic adherents to his new methods and discoveries, both from the classes of medical students of the Leipsic University and from the ranks of the profession, all of whom exhibited equal zeal and earnestness in developing the truths of this new school. From this nucleus, at Leipsic, this school steadily developed, materials and fabric being steadily brought forth by the new and energetic followers, for the building up and perfecting of the edifice of homœopathy. When to such yeoman service was added the experienced direction and masterly labors of Hahnemann, we may not wonder that the working of these new fields would produce an abundance of new and much needed materials. The suggestions of a practical experience, as well as the submitting of these materials to the careful examination of practical tests, would result in fast systematizing and giving shape to the form of polity of the School of Homœopathy.

(133)

The action of drugs upon the healthy system was the most prominent of the new branches of knowledge assiduously cultivated by the Homœopathic School; and this was so partly because of the hitherto totally neglected condition of this necessary department of medicine, and partly because, with the new system of treatment, so much stress was laid upon the absolute need of a knowledge of such action in order that the medicines might be used intelligently. Therefore, beside the further completion of the Materia Medica Pura, of which we have already spoken, many other treatises appeared upon this branch alone, in the literature of the new school within the next few years.

The position of the Homœopaths in relation to Allopathy, or Dogmatism, during this period, may be appreciated by some reference to the preface to Hahnemann's Materia Medica Pura. He says: " The day of a true knowledge of remedies, and a true system of therapeutics, will dawn, when physicians shall abandon the ridiculous method of mix‐ing together large portions of medicinal substances whose remedial virtues are only known experiment‐ally, or by vague praises, which is, in fact, not to know them at all." He goes on to say, further, that the theories which have hitherto swayed the minds of the profession must be abandoned, and that the principle that every single medical substance may aid to cure a disease by applying it, when indicated, in accordance to its homœopathic action must be examined.

He refers to the methods of a carpenter in his trade. Had a carpenter a piece of work to perform of which he had not a thorough understanding, were he ignorant of the kind of work for which each tool was best adapted, also had he not a correct knowledge of when to employ each tool, but arranging them together, impelled them at the work in question in a promiscous and indiscriminate manner, that would be compared to the method of allopathy.

Were he, instead, to acquire an intimate acquaintance with the character and uses of each tool, learning by observation which one would do the best work upon certain materials, and then, in their turn, as each was called for, employ these simply and skillfully, he often might use an insignificant or slender tool, yet one better adapted to the purpose, than if it were of an imposing and menacing aspect. This would seem on the other hand to be after the manner of the new school of medicine in its methods of shortening and overcoming disease.

Those reformers in the medical art, antecedent to Hahnemann, of whom Vesalius is a fair type, had, as we have seen, though with the effect to. bring upon themselves much personal rancor and malice from the representatives of Dogmatism, finally effected the full relinquishment on the part of Dogmatism of its opposition and persecution, and a full and complete adoption by this organization of those facts for which they had so devotedly stood and borne the brunt and shock of their onslaught.

Such men as Malpighi, Vesalius, and William Harvey, established certain facts triumphantly on behalf of the human race for all futurity, and now that the prejudice and malice in relation both to themselves and the cause for which they contended has entirely disappeared, their place in history and among the great discoverers is vouchsafed to them.

On the contrary, however, even at this day, and after a greater length of time than was necessary to accord to either of the above-mentioned benefactors of mankind, his place in history, we still find it too soon to give to Hahnemann the place which he shall occupy, because of the bitter prejudice and malevolence which still so prominently enter into a consideration of his name or work, or that of the school of which he was the founder.

Hence, the true merits and usefulness of his discoveries cannot be fairly and justly weighed, and there is need, in the case of Hahnemann, that a longer time transpire, than in the case of Vesalius; we can easily appreciate this, if we consider the different nature of his task.

In our day, and with the knowledge which has accumulated concerning the nature of mechanical force, we have learned not only to recognize the force of inertia, but also to know that the form under which force is used is an important consideration. For it is found that, if understanding the forms of work, a form of force be used, which can be utilized without loss, that the proposed effect is most successfully and economically produced. We find that our knowledge of motion, thus obtained, applies to the motion of animals as well as the motion of other bodies. If a body is to be raised to a small height we find it is most easily and therefore economically raised by the use of levers. In the use of the lever obtains the *weight* to be moved, the *power* to move it, and the *fulcrum*, or prop of the lever. We make a distinction of three kinds of levers, with particular reference to the position of the fulcrum. The third kind of lever, of which this figure is an example,

shows the power P between the weight W and the
fulcrum F. When we find that it is the provision
of nature, in the mechanism of man, to provide this
class of levers, we can see how wisely she has so
provided, for if the motive force, in the case of the
long bones of the leg, were brought to bear directly
on the resistance to be overcome, it would require
a large amount of motive force to perform what now
is, with such little motive force, easily produced.

Now we are coming to a consideration of this.
Supposing the lever (that is the bone) of the leg has
become broken in different places between the end
W and the point of the attachment of power, P. The
results will often be the raising of that part of the
lever still attached to the power, while the disunited
pieces will be found at varying angles and distances
from it. The remedy by readjusting these pieces
and keeping them in a position to make reunion
possible, is treatment which is of a mechanical
character, requiring the application of unyielding
forces equally along the broken lever, which shall
keep the broken fragments in exact apposition after
their replacement has been effected. If this illus-

tration is true of the human frame, as regards the employment of mechanics in treating wrongs to the system, it is also as true that chemicals are often needed as in poisoning and other maladies which require chemical helps. It was through such men as Vesalius and Harvey and Malpighi, with his microscope, that the proper direction to such knowledge was made possible in medicine, for, by the discovery of anatomical, physiological, and chemical facts, the nature of such occurrences could be correctly appreciated, and the proper chemical, mechanical, or surgical treatment offered.

What easy and simple tasks these compared to that life work which a previous state of ignorance imposed upon Hahnemann. For when the Galenists turned with rage to the works of Galen, and showed the heretic, Vesalius, where Galen emphatically said that the heart had an important bone in its anatomy, how easy for him to dissect heart after heart before them, and to demonstrate their error. Again, with Harvey, by actual demonstration, by vivisection, how easy to show the circulatory system to be a true hydraulic machine, with its pump, valves, and pipes, intended to force and carry the blood through the economy.

In comparison to these how herculean a task is
undertaken by Hahnemann! How difficult to demonstrate the departures from a state of health in
those insidious diseases which implicate those vital
organs, where either, because they are enclosed by a
bony box, as the brain, or lie deep within the other
viscera, examination is difficult and sometimes impossible, or where, on account of their inter-depen-
dence, a disturbance in the functions of one produces an unknowable quantity of malnutrition in
another! For an injury or disease in one organ,
irritating its nerves, produces (as is supposed) an
abnormal condition of the roots of these nerves
located in the spinal cord, and they, in turn, by
contiguity of surface, communicate this condition to
roots of adjacent nerves, through these, affecting a
vital organ, and producing serious disturbance
there, or, in a reflex way, disturbing the sympathetic
or other systems. If there are wrongs in the assimilative organs, or failures in the organs whose work
it is to carry off the waste and broken down tissues
of the system, or if there are certain deformities
which existed from birth, or from bad processes of
development, or from disease, as lesions or secondary
diseases, or if there are such factors as habit, toler-

ance, acclimation, or like varying external conditions at the time of examination, each has its individual effect, and renders the study most difficult and embarrassing.

These are illustrations of other more minute and complicated processes of diseases, which we must remember affect not one part alone of the body, and are not as easy of abatement or removal as is the weakened bolt with its worn and slipping thread, for the bolt can be repaired in such a manner as makes the illustration not analagous to the diseased vital organs of the human body. These are examples enough to show how difficult (if, indeed, it be not impossible) it is with the mutual and coexisting influences between the various tissues, organs, and systems of the human economy, together with their separate and often conjoint influence upon that very obscure and always most important factor, the vital force, to demonstrate, after the manner of the chemist or anatomist, the actual and absolute effect, or measure of interval traversed from disease and towards health, by the administration, at any time, of a given medicine.

Because, from the nature of the undertaking it was impossible to so demonstrate the absolute and unvarying condition produced upon the diseased organism, and as to repeated and renewed experiment to reproduce and continually load the absolute result into the floundering ship of Dogmatism, till to save and bring her safely into port would necessitate the throwing overboard of its ballast of prejudgment and prejudice; because in the nature of things this was impossible, therefore the Dogmatic sect still found it safe to assume the position towards this new curative method that these Homœopathic medicines were tasteless and without smell, and therefore without efficacy or power, and demanded that an absolute and unvarying effect by them be conclusively demonstrated.

Concerning the latter objections made, what better evidence has ever been brought forward of the appeal of any medical sect to those popular prejudices which sprung from the bias of the ignorant and unthinking masses? Who will attempt to show there is a necessary relation between the medicinal virtues of substances and their effect upon the taste and smell?

The answer brought forward to such objections was, that substances that are conceded to be poisonous to the human system have no taste and no smell;

therefore, are they without potency? Of the other strictures the Homœopaths did not attempt what from its nature was impossible, nor could they demonstrate what from its nature was unknowable. They have not tried to show absolute results as could Vesalius with the anatomy of the heart, but they could give results as compared to the results of Allopathic treatment as well as compared to their own from year to year, and decade to decade, that showed what the further development of their science was effecting in a practical way.

Again, they were met by the Dogmatic faculty with the statement that certain diseases appeared, and then, running a defined course, would, from their nature, exhaust themselves and disappear from the economy; also, in this connection it was said, that distressing symptoms would arise and vanish suddenly without outside interference, and they say because you cannot tell which symptoms were cured by homœopathic drugs, and which disappeared unassisted by treatment, therefore, homœopathic methods are unreliable and deceptive. To this, a partial answer was given to the effect that, this objection, if taken as asserted, it would apply equally to all systems and methods, and, if true, how could Allopathy claim to be efficacious? Why is not it, also, therefore, "unreliable and deceptive?" but we may more fully meet this question while answering the next objection.

The Dogmatists also said that the drug provings of homœopathy could not be reliable; how did they know but a small quantity of some other drug had been unconsciously taken at the same time, which would either produce the symptoms exhibited or antidote the one being proved? and, also, how could one discriminate between the true drug symptoms and imaginative ones? If it is not possible to show absolutely what are the effects of medicine and what not in the progress of disease, and if, by the simple method of Vesalius, this problem cannot, because of its character, be absolutely solved, homœopathy claims that it can, by comparative and inductive methods, solve this with certainty and correctness. The comparative and inductive method will carefully observe the action of a drug upon the healthy organism of many under differing circumstances, localities, and differences of age, sex, etc., and a careful record of these observations will show, by a comparison of them with each other, which are the symptoms of the imagination and which are those produced by other causes aside from the drug action, and also which are the prominent and significant symptoms of the drug. This same comparative and inductive method is then applied to the drug in testing it in the light of these provings at the bedside.

Let us go a little farther yet in this, that these two last objections may be fully met. When a homœopathic physician is treating certain diseases and symptoms and sees them disappear under the exhibition of a certain remedy, if the disease and other intercurring circumstances do not explain the alleviation of symptoms, and he is led to expect this effect from his knowledge of the characteristic action of the medicine, he assumes it as a curative effect of the drug, but not yet conclusively, until after continued experience in this disease he sees this ameliorating effect produced repeatedly by this indicated exhibition of this medicine. Or, if others report by the same comparative and inductive method the same results from exhibiting this drug under like circumstances then, in spite of the difficulties which nature has thrown in his way, he will announce with positiveness concerning the power

10

of this drug, under these homœopathic indications, to cure or abridge this disease.

Now, Homœopathy clearly shows that each drug has, upon the system, its own peculiar and elective action, this disease producing action being certain, and always, when experimented with under the same conditions, producing the same results—one drug, after being given to a healthy person, displaying, upon a localized surface in a certain portion of the economy, an eruption of a positive and peculiar character—another drug, when exhibited under like circumstances, producing, in a certain circumscribed portion of one lung, a pulmonic or inflammatory condition—another drug exerting an elective action upon the kidneys, producing disease while its influence is at work, which speedily disappears with a cessation of its use.

Despite these facts, as well as the additional one demonstrated by Homœopathy, that remedies should be used in the application of medicine in removing disease, that experiment proves to exert a specific action in health similar to this diseased condition under treatment—such facts, with their verification in the treatment of disease, having established the law as expressed by *similia similibus curantur—*

despite this, we say, Allopathy wholly ignores such properties and uses of drugs, declaring such methods to be false and harmful.

At that time, (1820,) Dogmatism having experienced, under the critical examination of Hahnemann, within the last twenty-five years, such complete discomfiture and rout, as regards its dogmas and doctrines, (such finished and perfect exposures of false and injurious methods in medical treatment, as the then accepted doctrine of Brunonism received at his hands, (1808,) being examples of his repeated and effective showing of the dangerous practices and theories employed and defended by this faculty,) renounced its old, and assumed a new position, peculiar as well as compromising to itself, although one which it was compelled from necessity to adopt, since there was nothing better or tenable presenting itself in this premise, nor has there been anything to this day.

It became apparent to the Dogmatic faculty that nothing had been productive of so much injury to this school as an organization, than these exposures by Hahnemann of the illustrations of the aptitudes and instincts which preëminently, since its inception, had influenced and ruled Dogmatism.

When, therefore, Dogmatism refused to become purged of this vicious nature, and to throw over "its immense commercial and corporation interests" in its old and established order of things, and accept the efforts of Hahnemann to reformation in medicine; and when later, at Hahnemann's hands, a rival school in medicine sprang forth which bade fair to give the future Dogmatic edicts in theory as rough handling as Dogmatism through Hahnemann had just received, there appeared this modification of its policy.

There was to be no loss of authority or assumptive power over the direction of thought, for this thing was to be subject to the indorsement of the Dogmatic faculty ; but it was the will and expression of Dogmatism, as concerning this question, that its adherents should use their own inclinations and follow their own predilections within certain limits; within these limits they could accept the theory of Brunonism, or Hoffman's, Sydenham's, or Stahl's hypotheses. Animism or expectant tenets or later cellular or neuro-pathological theories could be entertained singly and in their author's purity, or combined and amalgamated. By such fanciful theories, then, the adherents were to be guided in the treatment and management of disease. New and analagous theories strictly within these limits could also be developed subject to the indorsement of Dogmatism, but without these circumscribed limits the Allopath must not dare to direct himself, and the vigilance and penalties exerted to enforce the strict keeping of the faithful within these bounds, we shall presently, most carefully, and attentively consider.

This new state of things, however, was nothing but a summons by Dogmatism of certain of the prominent features of the Empiric School, upon which it had waged unceasing war in the time of the Ptolemies, but which it now desired to capture and appropriate for its own especial benefit and profit. In fact, then, Dogmatism emerges into view in this the nineteenth century as Empiricism; the adherents of this school so regarding this phase of matters with them, and for years it was so known.

This regard of the changes of dogma with the dominant school was in place here, for we had turned from a consideration of the homœopathic methods of treatment and of the allopathic objections thereto, and were to inquire, as the Allopathic School reject the truths affirmed by the Homœopathic School, as well as the methods of treatment to which they point, if the doctrines of Dogmatism and its treatment of disease would make a better showing, if the fire of argument, under which we have seen Homœopathy to stand unscathed, were to be turned upon the elder half-brother in medicine.

With the knowledge we have already acquired while standing upon the ever shifting sands which border upon the sea of Dogmatic fancy we shall not be surprised to learn what, through the eighteenth century, appears in the accepted methods of treatment by the dominant school.   First, there was used at this time a class of preparations—many hundred of them—called *antidotes*, many of which were a legacy from and may be taken as examples of the "assembled wisdom" bequeathed to Dogmatism by antiquity* (upon which it plumes itself so highly.) These preparations or *antidotes* were generally given in late years another term, so that they be not confounded with the class of chemical antidotes, they were therefore referred to as Theriaca.   This term is suggestive of molasses, and was probably applied because these preparations had its consistency and appearance.   We find one theriaca to be an electuary composed of one hundred and fifty different ingredients, and many others contained upwards of sixty different substances. Some of these antidotes were dismissed from the British Pharmacopœia in the middle of the eighteenth century, but others were prescribed up to and even later than the period with which we are now engaged, (1820.)

Some of the ingredients in these last are most loathsome, such, for instance, as VIPER'S FLESH, or vipers of a tawny color, which were to be boiled until the flesh dropped from the bones, and this to enter into a theriaca for administration in the case of some poor afflicted mortal who was so unfortunate as to fall into the hands of a physician of the dominant school; or, again, we might refer to the treatment of hydrophobia by a standard authority in Dogmatism, who, for this complaint, directed the keeping of the patient under water while the Psalm *miserere* was sung. We might point to him who was an authority even as late as the time of which we now write, (1820,) an authority than whom none among the shining lights of Dogmatism stood higher. In the treatises upon the practice of medicine, by Thomas Sydenham, we find that he directs the administration of such as the powder of stag's tail, bull's tail, boar's tooth. For pleurisy he prescribes

the bleeding of ten ounces of blood from the patient
and the giving of the "*expressed juice of horse dung*,"
also the "*spirit or volatile salt of man's blood, and of
vipers, and crab's eyes, etc.*," hog's lice was another
ingredient of the allopathic prescription.

Bleeding and drastic purgatives were the most
relied upon, however, in a very large class of diseases. The common and then well known saw,
" Physick, lies a bleeding," being intended to direct
the efforts of " Physick " into its widest and supposed
most useful channel. The use of both bleeding and
purgatives in these times will be appreciated when
we come to consider the Brunonian theories which
now influenced the direction of thought as applied
in selecting medical treatment. This system, holding most diseases to be a condition of sthenia, and
only to be cured by reducing the patient, the unfortunate was bled to fainting and syncope, and
drastic purges administered to effect the same result.

We may have an idea of the use of these methods
of Allopathy if we turn to the remark of Dr. Benja-
min Rush, a most eminent American physician,
whose medical treatises were as prominent and as
markedly able as any other of the medical works

which appeared with them in the first quarter of the nineteenth century. Dr. Rush declares that during his experience in medical practice, he has, in the bleeding of his patients, drawn enough blood to float a seventy-four gun man-of-war.

To the reader it may appear to be a difficult matter for the Allopaths to make such treatment as the above consistent with their law of *contraria*, the insignia which they fly at the mast head of their ship of Dogmatism, and point to as the concentrated essense of medical truth, the allusive figure of the assembled medical wisdom of the ages; but the reader may not be familiar with the elasticity of this wonderful law of *contraria*, or of the many incongruous shapes, the product of Dogmatic fancy, which this law has been called upon to cover in times past. The possibilities of this law in this direction will be apparent if we consider the methods of Dogmatism, and remember, also, that it is not possible for a treatment to be employed that will not present some phase or effect that is different or opposite in some of its particulars; how easy then for this medical school so prolific in theory and imaginative assumption to bring forward a new theory applicable to this treatment, which shall

give unusual prominence to this element of difference between the treatment and disease, and so at
once corroborate the application of the law of *contraria*, and substantiate the propriety of the Allopathic use of this treatment under these conditions.

The Allopathic school, then, in rejecting the new
methods of the Homœopathic school of dealing
with disease, as we have seen; in denouncing its
mode of preparing medicinal substances as both too
simple and unscientific to merit consideration, declaring that the very small quantity of a single
drug, when put into so much sugar of milk and
submitted to trituration, made it a conjectural
matter which part of this mixture contained the
drug, and, even if it were administered, would not
purge as it should, or produce emesis or diuresis, or
any other physiological effect, and, unless such effect
were produced, the disease could not be removed;
also, in declaring the Homœopathic law of *similia
similibus curantur* false, and warning the faithful of
the heresy of the teachings of the Homœopathic
school, and advising adherents of Dogmatism of
the necessity of the employment of discipline in its
ranks, as regards these heretical methods in particular, pronounced against each and all of these

methods and principles, and forbade the faithful to touch thereof.

In their stead the Dogmatic school had to offer the long prescriptions and filthy mixtures (of which those which we have noticed are extreme examples) and the use of the lancet in bleeding or of drastic drugs in purging, that the excessive depletion of the strength of the system by the use of one or both of these methods of robbing it of its flowing elements of strength might be accomplished, as a necessary means in the efforts of cure.

Aside from the fact that these, the prominent and characteristic methods of Dogmatism, more than carry their condemnation upon their face, we now see that the demurrer filed against Homœopathy by Dogmatism, and, without effect, applies with its full force against the ancient school of medicine. In fact, if we consider that a disease attacks a person, and being limited, eventually disappears of itself, what reason will Allopathy have to claim, that because it has reduced the patient by purging and vomiting, that this method of treatment has aided the cure? Surely any rational thinker will speedily conclude that it is nearer the truth to say that the patient recovered in *spite* of such treatment and not because of it.

In the arena of logic and argument the Homœo-pathic principles and law having, at every point, utterly vanquished Allopathy, we shall hereafter find the dominant school falling back almost wholly upon what has been to it before this time, its favorite auxiliary and indispensable reserved force, upon which it has often called to sustain the brunt and onslaught of Homœopathy, when not to receive such help would have been to suffer the most complete and utter rout of the foot and horse of Dogmatism.

This ally of which Allopathy has never lost sight, nor forgotten the consciousness of possessing, is called by name *mockery*, which in different phases we distinguish as *ridicule; mimicry; derision; contemptuous sneers, etc.* This ally, as we have said, Allopathy will fall back, and depend upon almost exclusively in the future, in the treatment of its rival in medicine, in argument, as well as in its persecution of Homœopathy in their mutual field of labor.

About this time a famous Austrian Field Marshal and Prince, sick, and given up as incurable by his physicians of the dominant school, repaired to Leipsic for treatment by Hahnemann, and while under his care died. Upon this, the charges were made on the part of the apothecaries, that Hahnemann had by his medicines caused the death of his patient, and by their influence or desperate measures, (ignoring the fact that none that they could produce could demonstrate a more thorough and exhaustive pharmaceutical and chemical knowledge than he,) finally succeeded in obtaining an injunction preventing Hahnemann from dispensing his own medicines. Instead of accepting the issue made here, and entering the lists, Hahnemann now, in his sixty-sixth year, longing for release from such strife, accepted the offer of physician in ordinary to the Prince of Coethen, an ardent admirer of his.

The enthusiastic coterie which surrounded Hahnemann and re-enforced him in his endeavors to advance the cause of Homœopathy were thus left without his able advice and direction to aid them in their work, except so far as he could, in a way, help them from a distance. The rich increase of homœopathic material within the next ten years, the result of the laborious study and research of the homœopathic co-workers was astounding, such men as Ernest Stapf, Fred. Ruckert, Frantz Hartman, G. W. Gross, Constantine Hering, Moritz Mueller, and many others contributing to this work. Most of these were physicians who were regularly educated and received into the Allopathic ranks and of acknowledged ability.

These materials appeared in a new edition of Hahnemann's "*Materia Medica*," and in five volumes of his "*Chronic Diseases*," in treatises upon Dietetics, manuals of Therapeutics, essays, and the Homœopathic periodical "*Archives*," first published in 1826; these, as well as many other papers, were from the pens of the more recent co-laborers in the School of Homœopathy.

From the year 1821, and through this decade, the new School of Medicine not only spread over Germany but was transplanted into every quarter of the civilized world. In 1821 it appears in Italy, in 1823 in Russia, and before 1830 we find it established and flourishing in France, Great Britain, and America.

From what we have seen in Germany we may not anticipate that the Allopathic profession evinced a willingness to accord those of the Homœopathic profession who appeared in these countries either fair or equitable treatment, and presently we shall take up this topic and endeavor, by a fair presentation of the facts, to get a true understanding concerning the history of the reception and difficulties, placed by the dominant school, in the way of the new school and its profession.

From 1830 marked differences arose between Hahnemann and others prominent in the Homœopathic profession, and as we shall have occasion to refer to this matter shortly, we will leave it here until the proper place is reached to show the cause and results of this difference in detail.

CHAPTER VII.

"OH! BEWARE, MY LORD, OF JEALOUSY; IT IS THE GREEN-
EYED MONSTER WHICH DOTH MOCK THE MEAT IT
FEEDS ON."

The afflux of visitors to Coethen from foreign
parts, drawn thither by the wonderful repute of
Hahnemann, and for the purpose of securing his
advice and skill, brought with it a beautiful and
wealthy young French lady, who made an impression
upon Hahnemann (now five years a widower) that
was fully reciprocated, and led to their engagement
and subsequent marriage, (1835.)

Shortly after this time he removed to Paris where
he spent the evening of his days, still engaged in
his professional and literary labors under most com-
fortable and felicitous circumstances. Eight years
after his removal to Paris, in the year 1843, and at
the age of eighty-nine, Hahnemann calmly and
peacefully passed from this life.

We may believe that the mockery and intolerance
which had persecuted and followed Hahnemann in
life, continued to assault his memory after his death.
One of the most prominent medical journals in

11   (161)

Great Britain and Ireland, the "*Dublin Medical Press*" said at this time: "It appears that old Hahnemann, the inventor of Homœopathy is dead, having prolonged his existence by infinitesimal doses of nothing to eighty-eight years, greatly to the consolalation and edification of the patrons and patronesses of quacks and quackery."

The old mockeries and criminations which had been so many times iterated and reiterated concerning Homœopathy, were now brought forward by the representatives of the dominant school with fresh zeal and virulence. The often repeated fiction, which to this day we hear recited, varying only in its application to different localities, that Homœopathy puts an ounce of epsom salts into the Thames, at Oxford, and expects a teaspoonful of the water at London to purge, was reproduced. Whereas, Homœopathy never pretended that its small doses produced a direct physiological effect upon the human system, as a poisonous and gross dose of a drug would, but always referred to the effect produced by the homœopathic triturations and doses of a drug, as being a manifestation of the rousing of or impression upon the *vital power* of the economy, and designated this effect of homœopathic doses, (as did Hahnemann before them,) the dynamic effect of medicine.

Again, the world was repeatedly informed that consistent Homœopathic treatment was to give the "hair of the dog for his bite." If a child fractures its skull by a fall upon the stony pavement the Homœopath (if he were not to compromise his professions) would cure it by again subjecting the injured caput to a collision with the same stones, for "like cures like, you know."

Concerning this, it seems useless to repeat that this is not in accordance with the law of similia but of identity. If one had a certain set of symptoms, and he had produced them with some poisonous chemical, the taking this chemical would be prescribing in accordance with the law of identity, for the identical substance is the self-same substance; if these symptoms, however, were caused by disease, and a drug, in homœopathic dose, administered, that was its nearest similar in its pathogenetic action; why this would be, together with the proper consideration of external conditions, in accordance with the law of similars, ("Like cures Like.")

Sometime in the decade of 1820, Wilhelm Lux advanced the theory of Isopathy, or the cure of disease by the reapplication of its cause, and advanced the law of *Æqualia æqualibus*, and we find that Allopathy gave some indorsement of this, the law of identity, by declaring that a prepared fox's liver would cure liver disease.

Again, we are sneeringly told that Homœopathy is a revelation in medicine, a faith; this is not true in any sense, for a revelation is a perfect and complete entity in itself, while Homœopathy is only a department of the field of medicine, (the therapeutics of the art of healing,) which was, in its infancy, gradually developed, imperfect, and full of mistakes, (which were, when discovered, promptly eradicated, as were the mesmeric assumptions, so soon after their entrance into the pale of the Homœopathic School,) but with the efforts of its scientific delvers and toilers in the field of experiment, has become more and more advanced and perfected, and commensurately more useful to suffering humanity.

Then, again, we have been, at intervals, constantly informed that *similia similibus curantur* is not true, and that Homœopathy is a humbug. Supposing the assertion "*similia similibus curantur* is not true" be assumed, then the question is, are the teachings of Allopathy and is the law of *contraria* any more true? Which is the nearest truth, and, consequently, which of the most use to man?

Again, it is charged that the only effect upon the
sick produced by homœopathic treatment is through
the imagination ; and this, because of the success in
the treatment of children's diseases, (which is fully
as great, proportionately, as with any other class of
complaints,) and because of the success in disease
where the imagination cannot enter as a factor, is
shown to be an untenable assertion.

How often we hear that the treatment of the
Homœopathic system is unscientific, because it is
neglectful, in that no attention is paid to dietetics
and hygiene, an assertion that can be shown to be
a misstatement by referring to the writings of
Hahnemann, from the last century forward. They
demonstrate to us how well he kept pace with his
time, both upon the subject of dietetics and upon
that of surrounding influences, and his papers "On
making the body hardy," "On dietetic conver-
sation," "Things that spoil the air," are pertinent
examples, as well as by the very extensive literature
of the Homœopathic school on Public and Private
Hygiene, Dietetics, and other branches of preventive
treatment.

In giving our attention now to the methods of
oppression and persecution which Allopathy has

always exerted towards the new school in medicine, we will, before leaving the recitals of occurrences in the German medical world, only state concerning it that the course towards Homœopathy here, with which we have become familiar, met with a temporary success, for, in or about the year 1830, and for the twelve succeeding years, governmental laws which, under Allopathic influence, were created and put in force forbade the Homœopathic physicians to prepare or dispense their medicines.

In or about the year 1827, Dr. Quin and two other physicians simultaneously appeared in England as adherents of the Homœopathic school, and practising its principles. At this time, and for thirty years thereafter, a physician, in England, was not considered to be duly qualified unless he possessed the diploma of the University of Oxford or of Cambridge, or that conferred by the Archbishop of Canterbury, and a license of the Royal College of Physicians of London, also the general practitioners must possess the diploma of the Society of Apothecaries.*

In 1833, Dr. Quin, having attracted public attention by his large and growing practice, was notified and admonished to cease practicing until he had

---

* A. C. Pope, M. D.

been duly examined and licensed under the common seal of the Royal College of Physicians.

Dr. Quin was not a fresh student, but an old practitioner at this time, and probably conscious, that in certain of the abstract and superfluous points of medical knowledge, which from lack of use in practical experience are speedily out of memory's reach, catch questions would be likely to be more than he would be equal to. He also felt the sensitiveness, which all physicians possess to a greater or less extent, of having laid himself liable to an imputation of not possessing the necessary and indispensable high order of intelligence requisite for the medical mind. In other words, he feared that because he was a Homœopath, therefore, he would not receive the equal and fair treatment that his Allopathic brethren would in their examination before this board.

For this he had the courage to defy their injunctions, and to dare to continue to practice despite their mandate and admonition. It may be observed in this connection that the demand upon Dr. Quin to appear before the examining board was nothing more than was required of every practicing Allopath " within seven miles of London," but we shall see if his apprehensions, relating to this question were not substantiated by medical events in the near future.

An element which asserted itself about this time, and thenceforward, an element which is often most conspicuous and always influential, an element over which the allocutions of the Allopathic faculty produce no apparent result, and therefore has sorely tried and embarrassed the dominant sectators, is the clientage or laity of Homœopathy. After examining, from a practical standpoint, the merits of the new school of medicine, accepting its claims, and declaring in favor of trusting to its competency, as concerned themselves and their own, where the offices of medicine were required, this class has ever exhibited an inclination to uphold the hands of the Homœopathic profession in its unequal fight for a fair and equitable consideration before the law and the public.

It is to the opposition and influence of this element against the efforts of the Allopathic faculty to create laws in governmental bodies, which were to vest in the Allopathic faculty full powers in affairs of medicine, or to create laws intended to directly proscribe the Homœopathic profession, and by a persecution similar to that of Hahnemann's time discourage and blot out this healthy rival of Dogmatism; it is to this class, indeed, that Homœopathy owes, in a great measure, that its opportunity for advancement and progress in future was made possible.

The attempts made in Parliament to enact such laws as that presented in 1858, which vested it with a proposed Allopathic medical council to decide if any duly qualified practitioner were practicing "irregularly," and if so, to erase him from the register, and so, by the application of the law, compel him to desist from practicing his profession, were frustrated, however, by the hands of the Homœopathic laity members of this house.

In fact, the activity of many men most prominent in their profession or business, and influential in municipal and governmental affairs, in advancing the interest of Homœopathy and in identifying themselves with its undertakings, was no doubt a reason for instigating the Allopathic profession to prosecute its animosity to the new school with renewed vigor.

The success and popularity achieved by the Homœopathic physicians in London were duplicated in the other large cities of Great Britain. Dr. Black and Dr. Russell, in Edinburgh, by their large and

growing reputation, called attention to the claims of
the new school, and soon the ire and horror of the
faculty of the celebrated medical school here was
provoked by the announcement of Dr. Henderson,
the able and celebrated professor of Pathology at
this University, that he had investigated and accepted the principles of the new school of medicine.
For this reason and upon this ground the effort was
made under the leadership of Dr. Syme, his colleague,
to dispossess him of his position in the faculty, and
it was finally successful.

Shortly after the Medico-Chirurgical Society of
Edinburgh (the Allopathic society here) passed this
resolution, " that the public profession of Homœopathy shall be held to disqualify for being admitted
or remaining a member," and immediately thereafter expelled Dr. Henderson from the society.

Shorty after this Mr. Pope having followed the
prescribed course of medicine at this University,
regularly and in due time presented himself to the
board of examiners for examination for graduation
and his degree in medicine. It was made known
that he had been investigating the teachings of the
new school, and upon concluding his examination,
he was asked if he was intending to espouse Homœo-

pathic principles, and upon his answering, that at present he could only say that as yet he was only investigating this system of medicine; his case was carried before the faculty, the outcome of it all being that Mr. Pope was compelled to leave the University without his diploma, not because of a failure on his part to comply with any of the requirements, or because his examination was not satisfactory in every way; the reason of the refusal of the faculty to grant to him the degree of the University being, that he would not bind himself in advance of his investigation of the subject *not to favor or practice Homœopathy.*

Soon after this, the University of Aberdeen, St. Andrew's, as well as other universities in Great Britain, indorsed the position of this University of Edinburgh, and resolved to take the same ground, and thereafter declined to permit candidates for examination to appear before them if they refused to renounce the principles of Homœopathy.

The court of examiners of the Society of Apothecaries, a society whose diploma was essential to those medical practitioners who practiced medicine under the law as we have previously shown, declared that "in their capacity of examiners they will refuse their certificate to any candidate who proposes during the examination to found his practice on what are called Homœopathic principles." Later, the secretary in reply to an inquiry in 1853, answered:

'SIR: The Court of Examiners still refuses to admit any person who calls himself a Homœopathist.

" I am, sir, yours, etc.,

" H. BLATCH, *Secretary.*"

Lastly, if we turn to the " *Provincial Medical and Surgical Association,*" and study its action upon this question, and also pay particular attention relative to the bearing and influence of this association upon medical matters, we shall gain more insight than ever into the organization and true methods of Dogmatism controlling the affairs pertaining to and affecting the realm of medicine.

This association was first formed in the year 1833, and was intended to have jurisdiction in Great Britain outside of London. Its first announcements and meetings proclaimed that it would be composed of members, the most prominent physicians in the several provincial towns.

This association was created that Dogmatism might live, its laws and code were to be the accepted law and faith of each and every institution and individual who, in his profession, desired to be identified with this school. The laws of this association, which were intended to govern the members of the Allopathic profession, were gathered together into a code, and received the title of the "Code of Ethics of the Medical and Surgical Association!"

Henceforth let no individual adherent of Allopathy, nor any institution as a medical college, medical society, or other representative body of Dogmatic medicine, dare to disregard or break this code of ethics. For, as it was to this association, that each and every of the above mentioned looked for indorsement as to their orthodoxy in medicine, so, therefore, they must not only leave it to this association to decide for them what shall be their course of action in professional and medical matters, as well as what is the right and what the wrong thing from a moral point of view, but they must so thoroughly understand and carefully follow this code of ethics that, whether they be college, society, or individual, they suffer not the excommunication of this association to be uttered against them, else those now their associates and brothers in the faith will not approach or have professional intercourse with them, lest, by so doing, the association would declare their garments to be defiled also.

We understand now why this unanimous and harmonious action which was, and is always, explained, as for the good of Dogmatism, appeared, and to show that this spirit, as well as this *modus operandi*, from and after this time, was not peculiar to this quarter of the world but extended to most distant portions also, we will, after considering the action of this association in relation to the new school of medicine, pay some attention to the medical arena in America.

At the meeting of this, "The Provincial Medical and Surgical Association of Great Britain," at Brighton, in 1851, these resolutions were presented and passed: "First, That it is the opinion of this Association that Homœopathy, as propounded by Hahnemann and practiced by his followers, is so utterly opposed to science and common sense, as well as so completely at variance with the experience of the medical profession, that it ought to be in no way or degree practiced or countenanced by any regularly educated medical practitioner. Second and third, That as Homœopathists have spoken contemptuously of medicine as regularly practiced, it is derogatory to the honor of members of this Association to hold any kind of professional intercourse with them. Fourth, *That real Homœopathic practitioners, those who practice Homœopathy in combination with other systems of treatment, and those who hold professional intercourse with Homœopathists ought not to be members of this Association."*

We wish also to append these resolutions, passed in 1826 by the College of Surgeons of Ireland: "That no fellow or licentiate of the Royal College shall pretend or profess to cure disease by the deception called Homœopathy, or the practice called Mesmerism, or by any other form of quackery. It is also hereby ordained that no fellow or licentiate of the College shall consult with, meet, advise, direct, or assist any person engaged in such deceptions or practices, or in any system or practice considered derogatory or dishonorable by physicians or surgeons."

Although in the decade of 1820, Dr. Gram introduced Homœopathy into the United States by settling in New York city, to there enter into the practice of his profession, yet for some years the persecution and ridicule which was visited upon him, as well as upon those who followed him and shortly afterward represented the new school in this and its neighboring States, was a localized effort on the part of Allopathy, and so continued until 1847.

The personal enmity exhibited by the Allopathic profession towards Dr. Gram, a man of spotless honor and of unexceptional qualifications for those times, (having been educated in one of the prominent universities in Europe, and there taking one of the highest degrees,) is well established and a matter of common knowledge. After this time, (1847,) however, we find the methods of oppression and mockery which Allopathy persistently pursues towards the new school of medicine in the United States, being concordantly employed in all parts of the country, and under the direction of an acknowledged controling power.

From this time a National Medical Society asserts itself, known as the American Medical Association, and since we have learned of the intent and scope of the Provincial Medical and Surgical Association of Great Britain, we shall be interested to learn of the position and bent of this society as regards the younger half-brother in medicine.

We find this Association to be composed of members who are delegates from a recognized Allopathic medical society or college or like institution and of medical practitioners in " regular standing " who have received the unanimous vote of the Association.

We find their term "regular practitioner" defined to their satisfaction in this portion of their "Code of Ethics," "no one can be considered as a regular practitioner or a fit associate in consultation, whose practice is based on an *exclusive dogma*," etc. We may better understand the aims and drift of this organization, and its intent to duplicate in America the position, power, and place occupied by the Provincial Medical and Surgical Association in Great Britain, if we give attention to these declared acts created by the American Medical Association for the purpose of guiding and governing the Allopathic profession in the United States.

" *Resolved,* That no State or local society shall hereafter be entitled to representation in this Association that has not accepted its Code of Ethics.

" *Resolved,* That no State or local society that has intentionally violated or disregarded any article or clause in the Code of Ethics shall any longer be entitled to representation in this body.

" *Resolved,* That no organization or institution shall be considered in good standing which has not adopted its (The American Medical Association) Code of Ethics."

We see, therefore, that the power of disciplining and of governing every individual member of the Allopathic profession in this country, is as absolute with this National Allopathic Society as it is with

12

its archetype in Great Britain, with which we have become somewhat acquainted.

For if every practicing physician (having previously followed the prescribed curriculum of medical study and being regularly graduated in medicine) applies to his local medical society for recognition and membership, which a proper personal examination and thorough scrutiny of his credentials should secure, by the evidence of this connection with this local society, his standing as a medical practitioner is placed before his neighbors and beyond cavil, and, therefore, himself distinguished from the ignorant and unqualified pretenders to a knowledge of the curative art, who are found in every locality.

It was also of equal importance that the several medical societies be duly and generally recognized as the proper exponents of this function and as representative of the dominant school in their premises.

This may also be found to apply with equal force to the several medical colleges, for if it should appear that such an Allopathic College was under the ban, as regards the supreme authority in orthodox medicine in the country, how very materially would it affect the number of students seeking admission at its doors when it was appreciated by the medical world, that this ban would prevent any Allopathic medical society or physician from recognizing the virtue of the diploma of this college?

Hence, the dicta of the American Medical Association, as represented in its Code of Ethics, was duly acknowledged and adopted by each and every medical society and college in the country and its tenets carefully observed. We shall better understand the absolute authority of this supreme council if we follow it in an instance of its application of its method of discipline.

In the year 1870 the American Medical Association held its meeting at Washington, D. C., and among the other regularly chosen delegates which assembled there to participate in this meeting were the delegates from the Massachusetts Medical Society, the recognized Allopathic State Society of Massachusetts. These delegates received their requisite credentials from the Massachusetts Society, and presented them in due form and at the proper time. When the American Medical Association took action towards these delegates, though premeditated by the association, yet wholly without notice as regards the Massachusetts Medical Society, it was, both to its delegates and itself, a complete surprise.

This Association refused the right of the delegates of the Massachusetts Medical Society to sit in this meeting of the American Medical Association until their right to appear as delegates should be determined.

Later, after investigation, the Association finds " that the Massachusetts Medical Society voluntarily and improperly furnishes shelter and gives countenance to irregular practitioners, to such extent as to render it unworthy of representation in the General Assembly of American Physicians." In other words, the Massachusetts Medical Society had fallen from grace, and no longer represented true orthodoxy in medicine in the State of Massachusetts.

Let us see if this discipline has its intended effect. Although the expression, " irregular practitioner," is rather indefinite and lax, there was not here any doubt of its intended application; in fact, we have already seen it to be the term which is most commonly used to designate the Homœopathic practitioners, and to be synonymous, in this connection, with " quack," " charlatan," and the like, in their application to the new school of medicine.

In consequence of this action of the Supreme Allopathic Council, the Massachusetts Medical So-

ciety, at its next meeting, in the succeeding year,
(1871,)

" *Resolved*, That if any fellow of the Massachusetts Medical
Society shall be or shall become a member of any society which
adopts as its principle in the treatment of disease any exclusive
theory or dogma, (as, for example, those specified in Article 1 of
the By-Laws of this Society,) or himself shall practice or profess
to practice, or shall aid or abet any person or persons practicing
or professing to practice according to any such theory or dogma,
he shall be deemed to have violated the By-Laws of the Massachusetts Medical Society by '' conduct unbecoming and unworthy
an honorable physician and member of this Society.''—By-Laws,
VII, 5.

" *Resolved*, In case the society concur with the councillors in
the foregoing resolution, that the President of the Society shall
appoint a committee of five fellows (to hold office one year, and
until others are appointed) to bring before a board of trial any
fellow who, three months from this date or after, shall be found
chargeable with the offense set forth in the foregoing resolution.''

Against whom these resolutions were directed will
be readily understood if we know that there were at
this time several members of the Massachusetts
Medical Society who had been admitted as members
thereto, yet after finishing the course of study at the
medical school of Harvard University, (Allopathic,)
and entering upon the practice of medicine as a
true and orthodox Allopath, had become disgusted
and discouraged with the results of Allopathic
methods, and were led to investigate the induce-

ments which the success of Homœopathic practice evidenced.

These persons had recognized the truth of the principles of the new school of medicine, and avowed as much, as well as a determination to use these in the sick room where they would establish better results. That these physicians were educated after a manner that would be most gratifying to the Allopathic sense, must be acknowledged when we know that this is certified to by the diploma of the first Allopathic medical school in this country, and that they were the physicians at whom these resolutions were aimed, will be more fully appreciated when we quote the resolution which was passed before the above, but probably because it was found to be inadequate to the purpose was followed by those which we have just read.

This resolution ran: " *Resolved,* That the Massachusetts Medical Society hereby expels from fellowship all those who publicly profess to practice in accordance with any exclusive dogma, whether calling themselves Homœopaths, Hydropaths, Eclectics, or what not, in violation of the Code of Ethics of the American Medical Association."

Those summoned by virtue of the action of the

Massachusetts Medical Society, as set forth in the
before quoted resolutions, were each and all physicians who were known to be admitted Homœopaths.
The summons read as follows :

"NORTHAMPTON, MASS., *November* 4, 1871.
" To —— ——, M. D.,

"SIR: Charges having been preferred against you by a committee of the Massachusetts Medical Society of " conduct unbecoming and unworthy an honorable physician and member of this
Society," to wit : " by practicing, or professing to practice, according to an exclusive theory or dogma, and by belonging to a society whose purpose is at variance with the principles of, and tends
to disorganize the Massachusetts Medical Society."

" You are hereby directed to appear before a board of trial at the
Society's rooms, No. 36 Temple Place, Perkins Building, on Tues-
day, November 21, 1871, at 11 o'clock a. m., to answer to the
same, in accordance with the by-laws and instructions of the
Society.

" SAMUEL A. FISK,
" *President of the Massachusetts Medical Society.*"

The gentlemen thus summoned obtained a temporary injunction from the Supreme Court which compelled adjournment of the society for a time. The
charges and specifications which the board were
compelled to bring forward when the temporary
injunction disappeared, were as follows :

" The committee now specify that the exclusive theory or dogma
referred to in said charges, is the theory or dogma known as Homœopathy, and the Society therein referred to, whose purpose is at
variance with, and tends to disorganize, the Massachusetts Medical Society, is the Massachusetts Homœopathic Medical Society.

" The committee file the following as further specifications :

" CHARGE I.—That you are guilty of an attempt to disorganize, and destroy the Massachusetts Medical Society.

" SPECIFICATION 1.—That you have joined, and are a member of, a certain society, known as the Massachusetts Homœopathic Medical Society, whose purposes are at variance with, and which tends to disorganize, the Massachusetts Medical Society.

" SPECIFICATION 2.—That you belong to, and are a member of, a certain society called the Massachusetts Homœopathic Medical Society, which adopts as its principle, in the treatment of disease, a certain exclusive theory or dogma known as Homœopathy.

" CHARGE II.—That you are guilty of conduct unbecoming and unworthy an honorable physician and member of the Massachusetts Medical Society.

" SPECIFICATION 1.—In that you practice, or profess to practice, medicine according to a certain exclusive theory or dogma known as Homœopathy.

" SPECIFICATION 2.—In that, while a member of the Massachusetts Medical Society, you have joined, and are a member of, a certain society called the Massachusetts Homœopathic Medical Society, which adopts as its principle, in the treatment of disease, a certain exclusive theory or dogma known as Homœopathy, and whose purposes are at variance with, and which tends to disorganize, the Massachusetts Medical Society.

" SPECIFICATION 3.—In that you are a member of a certain society, called the Massachusetts Homœopathic Medical Society, which adopts, as its principle in the treatment of disease, a certain exclusive theory or dogma, known as Homœopathy, whose purposes are at variance with, and which tends to disorganize the Massachusetts Medical Society.

" You are hereby further reminded that, to try the same, the Board of Trial *stands adjourned* to April 29th, 1873, at 11 A. M., at 36 Temple Place.

" GEORGE C. SHATTUCK,

" *President of the Massachusetts Medical Society.*"

The manner of conducting the trial is exhibited by the action taken on the following demands made by the Board of Trial :

" 1. That the trial should not be held with closed doors, but that their friends should be allowed to be present.

" Demand refused.

" 2. That reporters for the press should be allowed to be present; that as this was a matter affecting the character of the accused, the public had a right to know the evidence produced, and the manner of conducting this trial.

" Demand refused.

" 3. That the accused be allowed legal counsel, since it is proposed to dispossess them of rights, privileges, and personal property.

" Demand refused.

" 4. That they be allowed to have an advocate, not a member of the Massachusetts Medical Society, present to advise them.

" Demand refused.

" 5. That, as they have reason to object to the record of the Secretary, a phonographic reporter of the trial should be appointed by mutual consent, and sworn to the faithful performance of his duty.

" Demand refused.

" 6. That the accused may employ a phonographic reporter.

" Demand refused.

" 7. That an amanuensis, not a member of the Massachusetts Medical Society, be allowed to sit beside the accused and assist him in taking notes of the trial.

" Demand refused.

" 8. The right to peremptory challenge.

Demand refused.

" 9. The right to challenge members of the Board of Trial for good and sufficient reasons.

" Demand refused." *

---

* Pages 1015–'17–'18, Transactions American Institute of Homœopathy, 1876, Vol. II.

Of course we do not need to be informed that the result of this was anticipated and arranged before its commencement, for we know that its inception and institution was because of the edict of the American Medical Association in these premises, and such direction was not issued that a fair trial of those alleged offenders should obtain, but that they should be expelled and dislodged from the Massachusetts Medical Society.

If, after carefully reading these methods of discipline, asserted by the Supreme Council of Dogmatism, "in the middle of the afternoon of the nineteenth century," any one should infer that the sole reason of such action and course towards the new school was the thorough rout and extinguishment of this school, they should here correct themselves, for the policy and course of the dominant school renders the belief necessary that more often of late years those subsidiary interests, which are looked upon by the Allopathic faculty as of so much importance, were also considered, and because the old and usual oppressive policy of Dogmatism would here exert immediate results to the profit of the dominant school in this direction, therefore was it continued and energetically prosecuted.

Let us point this with a good illustration. We now can understand that not only is the Allopathic physician prevented by his accepted code from consulting with a homœopath, for full well he knows that if such code be departed from by his own act that his local society must either expel him or else be in danger of being itself put under the ban by the State or National Society, but we shall see that this policy of the dominant school will assert itself in full force in the charitable institutions. We know that at this day those humane and necessary institutions, the public hospitals, in this country, aggregate, in their first cost, millions upon millions of dollars, which was, of course, met by drafts upon the public purse, (a public, by the way, that has a large and rapidly increasing minority, influential and wealthy, who are pronounced in favor of employing the method of the new school of medicine,) and when we consider the prominence which those having the medical management of these institutions must acquire, and the reputation and benefits which those physicians must reap, because of their occupying a position which carries the care and responsibility of so many and varied diseases, we must at once recognize the immense advantage to be acquired by any school of medicine over other schools, by the complete monopoly of the management of public hospitals; for such a position, under honest and economical administrations, even, must result in the placing of many and important favors to the credit of the controlling school and of the linking of their interests, to a greater or less extent, with that of other and influential corporeal bodies.

The Allopathic school has, by insisting that here its Code of Ethics must enter, and with it rule supreme, declined to, in any way, permit the new school to conjointly occupy the public hospitals, in the sense that, if one be carried to such public institution who has been accustomed to and prefers the attention and care of a physician who uses homœopathic methods in his treatment, and though such a patient be accepting the hospitality which the free wards here offer, under no circumstances can such patient (even though we find this hospital to be supported by a tax-paying public, one-fourth of whom are pronouncedly and always for placing themselves and their own under the care of the physicians of the new school) presume to call in and advise with a physician of the new school, though such request as regards any Allopathic physician in regular standing would not be refused.

So successfully is this policy of the effectual exclusion of Homœopathic physicians from public institutions as physicians executed, and thus a full, fair, and continued comparison of the results of the treatment of the different schools in the hospitals, that mankind, in general, might profit by such comparison prevented, that we find the medical branch of the Federal army in the late civil war so thoroughly and efficiently managed in the interest of the dominant school that, in no instance, did a Homœopathic surgeon succeed in entering the army as such, the very few that did succeed in obtaining appointments as surgeons succeeded because they presented their diplomas obtained from a regular Allopathic college, and, while in the army, acted with and comported themselves as Allopaths, so far as could be discovered.

We may not charge this to a lack of patriotism upon the part of the Homœopathic surgeons, for many not succeeding in obtaining appointments entered the army as privates; numbers upon numbers of Homœopathic surgeons so doing, and subsequently declaring that the fear of their possible subjection to the tender mercies of the representatives of Allopathy being far greater to them than that of the other fortunes of war.

We shall speedily recognize the possibilities of such complete success if we remember that a surgeon in the army has an absolute military rank, and that his functions cannot be questioned by officers in other branches of service,* and also that the board which has final and full powers to act upon all applications for position in the medical corps of the army, was chosen by those representatives of the Allopathic faculty then in place upon this corps, and that such choice was made from members of this corps solely, it would seem hardly necessary to add.

Extended illustrations of the experiences of Homœopathic surgeons while endeavoring to secure positions upon the medical corps of the army at the beginning of the civil war, are to be met with in the literature of the Homœopathic profession in the several States ; those interested in acquainting themselves with such examples are directed to the second volume of New York State Homœopathic Transactions, page 208, which gives an account of the application of Dr. T. D. Stow, which was filed by him in 1861, and though filed in due form, and

---

* The Surgeon General has rank of Brigadier General of the Army.

fulfilling all the requirements of the law, was yet rejected *because he was a Homœopath*.

With this following of the course of action and bearing of the dominant school towards its rival in medicine, in which is shown its efforts to cause legislation, which would place the infant half-brother in medicine where the hand of his elder could asphyxiate his more helpless relation, we shall be ready to appreciate the needed assistance to the cause of the new school, which was rendered when the element known as the laity entered the lists to defeat the designs of Dogmatism here, but since this time it has, in a field where the laity could not follow to take part, persistently and ruth-lessly used ever effort and every unfair means to subvert the true aims of medicine, if only the young rival of Dogmatism could be crippled and the interests of Dogmatism advanced.

> Believe, as we believe, no more, no less,
> That we are right, and nothing less confess;
> By the Code of Ethics and its mandates we abide
> And concede such other things as with it coincide.
> Think ye as we think, and do as we do,
> And then, and only then, we'll fellowship with you.
>
> That we are right, and always right, we know,
> For the "assembled wisdom of the ages" tells us so,

And to be right is simply this to be
Entirely and in all respects as we
To deviate a hair's breadth or begin
To question, or to doubt, or hesitate, is sin.

'T were better that the sick should die than live,
Unless they take the medicine we give;
Let sink the drowning if he will not swim
Upon the plank that we throw out to him;
'T were better that the world stand still than move
In any other way than that which we approve.

CHAPTER VIII.

With this showing of the predilections and bent
of the sectators of the dominant school, which we
see equally well illustrated in their determination
to remain secure and unmolested while they "sat by
the flesh-pots," as well as by their oppressive and
uncharitable animadversions, which they endeavored
to obtrude upon the new school of medicine, we will
now try to more fully appreciate by a careful and
impartial consideration the charges which they lay
at the door of the new school, charges which we
have seen to abound in accusations, assumptions,
and pretexts, and brought forward by them to justify
their course of action.

The quotations which have just been given were
shown to be the acts of representative bodies and
societies of the dominant school of medicine. In
these several quotations the assumptions, charges,
and pretexts are found to consist of—First, That
Homœopathy is opposed to science and common

13   (193)

sense, and completely at variance with the experience of the medical profession. Second, That Homœopathists have spoken contemptuously of medicine as regularly practiced. Third, That it is derogatory to the honor of members of the Association to hold any kind of professional intercourse with Homœopaths. Fourth, That the "deception called Homœopathy," is quackery, and that no member should "consult with, advise, direct, or assist any person practicing Homœopathy." Fifth, That Homœopathy is an exclusive dogma. Sixth, That no one can be considered as a regular practitioner or fit associate in consultation whose practice is based on an exclusive dogma. Seventh, An adherence to the Homœopathic practice is conduct unbecoming and unworthy an honorable physician. Eighth, "In that you are a member of a certain society called the Massachusetts Homœopathic Medical Society, which adopts as its principle in the treatment of disease a certain exclusive theory or dogma known as Homœopathy, *whose purposes are at variance with and which tends to disorganize the Massachusetts Medical Society.*"

This last is a quotation from the charges made by the Massachusetts Medical Society, and upon which those members on trial were judged and prejudged and therefore expelled from this society, and we wish to direct particular attention to the spirit of self-seeking here exhibited.

For what was the art of medicine instituted, and why is it maintained ? For the self-aggrandizement of some body of a few hundred who avow themselves (as regards a population of hundreds of thousands for whom they assume to act,) to be a close corporation, and the custodian and sole receptacle of all medical knowledge ! Is it for the mercenary aims and purposes of these few that medical science is fostered ? The purpose of Homœopathy is at variance with and tends to disorganize the Massachusetts Medical Society, and, therefore, is deserving of the anathema and obloquy which all truly good Allopaths should heap upon it. Avaunt with such inanity : withhold your slanders and misrepresentations, and show which is the most efficacious method of treating disease, and which the practice that will best serve humanity in the safest, speediest, and most pleasant manner; give the world some of your results after investigating Homœopathy, and what have been the comparative results in the treatment of disease by the two schools,

and see if you can from this show why you should
inveigh against the new method of healing and
not accord to the infant, but promising child, of
medicine a " fair field and no favor."

The reader, however, shall not be deprived of the
privilege of viewing such comparison. Dr. Routh,
a distinguished Allopathic physician, in a treatise
entitled the " Fallacies of Homœopathy," gives these
statistics of the comparative results in the Vienna
hospitals in

HOMŒOPATHIC TREATMENT.	ALLOPATHIC TREATMENT.
Deaths, per cent.	*Deaths, per cent.*
Pneumonia . 5.7	24
Pleuritis . 3	. 13
Peritonitis . 4	. 13
Dysentery 3	22

Dr. Routh tries to explain this showing by claiming that more severe cases presented themselves at
the Allopathic hospitals than at the Homœopathic,
but an eminent Allopathic physician, Dr. Wylde,
asserts, concerning this point, that " the cases he saw
treated at Dr. Fleischman's Homœopathic Hospital
were fully as acute and virulent as any he had
observed elsewhere."

In 1855 the House of Commons ordered printed

the comparative death rate during the malignant epidemic of Asiatic cholera in 1854, and this report showed:

Asiatic Cholera—Deaths under Homœopathy, 16.4 per cent.; Allopathy, 59.2 per cent.

Here is a carefully and correctly prepared table of death rates in 1870, 1871, and 1872, in the private practice of physicians of the city of Boston, Massachusetts.

Year.	ALLOPATHIC.		
	No. of Physicians.	No. of Deaths.	Average deaths to each Physician.
1870	218	3,872	17.76
1871	233	3,369	14.46
1872	233	4,575	19.63
Total	684	11,816	17.27
	HOMŒOPATHIC.		
1870	40	402	10.05
1871	44	363	8.25
1872	54	446	8.26
Total	138	1,211	8.77

A question to be considered here is, were the Homœopathic physicians having as large an average practice as the Allopaths? We leave the answer to each reader. Are the Homœopathic physicians in your own city, as a rule, more or less busy than the average Allopath?

An insurance company, the Homœopathic Life Insurance Company of New York, (claiming that, as a business venture, there is greater profit in insuring the lives of those who employ homœopathic treatment in disease, and having now years of experience and facts to indorse its statement, that they point to the report of the Insurance Commissioners of New York, which give the rate of mortality of this company as less than the average of insurance companies, hence, therefore, its peculiar feature of less rates to Homœopaths is grounded in permanence,) furnishes the following statement over the signature of Dr. J. B. Tuttle, a physician to the Michigan State Prison for years:

	Average No. of convicts per annum.	Total No. of deaths.	Total No. of days labor lost.	Total cost of hospital stores.
Under allopathic treatment in 1857, 1858 and 1859 ____	435	39	23,000	$1,678
Under homœopathic treatment in 1860, 1861 and 1862 _____	545	20	10,000	500

"This improvement was obtained, notwithstanding I had to contend, during the years 1861 and 1862, with epidemics of smallpox, of which there were thirty-two cases; of measles, of which there were thirty cases; and of sporadic cholera, of which there were forty-four cases. Many of these latter were of a very severe type; but all were successfully treated, and speedily cured, by infinitesimal doses, and without any resort to any kind of 'heroic medication.'

"And here I may remark that the success of the Homœopathic treatment was so great that many of its opponents attempted to account for it in other than the right and legitimate way. They affirmed that the good health of the inmates of the prison was owing entirely to the abundant supply of pure artesian water which had been introduced a short time previous to my appointment. But they failed to see that the water lost its efficacy soon after Homœopathic practice was abandoned, and that it did not regain its virtues until that system was again adopted in 1872; all of which may be seen by referring to the Prison Reports during the ten years when Allopathy was 'in' and Homœopathy was 'out.'

"Taking another and later comparison, we find that, in round numbers:

	Days labor lost by sickness.	Cost of hospital stores.
Under Allopathic treatment in 1870, '71..	24,000	$1,800
Under Homœopathic treatment in 1873, '74.	11,000	900

" While the average number of convicts during the last two years was greater than ever before in the history of the prison."

These comparative tables are produced here for the reader's benefit, and not to meet the question for the Massachusetts Medical Society, for, as we have seen, they ignore such facts, and decline to concern themselves with them.

Those charges against the new school that have been here numbered three, four, six, and seven, respectively, are of a nature impossible to offer answer by argument. They are, each and all, assertions from authority, and, as far as can be seen, without basis of fact, or evidence that there are existing reasons springing from a scrutiny or examination of the subject; they are, indeed, possible only because the position of authority of the allocution was such as to render such authority competent

to assert that " it was derogatory to the honor " " for
a regular practitioner to associate," etc.

These assumptions rest so exclusively upon authority alone that we will find them to apply equally
to any other subject, for instance, why would it not
read with equal force thus. " The deception called
the telephone is quackery, and no member shall be
considered a fit associate in consultation who resorts
to, or is in the practice of, using this shallow delu-
sion." " The acknowledged and persistent use of
the telephone is conduct unbecoming and unworthy
an honorable physician."

What exception could the regulation Allopath,
who finds it best to harbor and use telephones in
his office and business, make to the above were it
brought against him by the act of the Supreme
Allopathic Council, that would not, with equal force,
be found to hold equally well for the poor persecuted Homœopath in his efforts for the establishment of the Homœopathic principles.

The charge that Homœopathy is an " exclusive
dogma " is, as we have seen, the most frequently
asserted one. It refers of course to its law of similia,
and we are continually confronted with the assertion
that it is claimed for this law that it is " exclusive "
and " universal."

That it is an exclusive and universal law we contend, but no doubt the Allopath, with his reserve
of invective, will find that the significance by him
attached to the expression " exclusive dogma," and
exclusive or universal law, is neither shared by the
Homœopathic profession nor acknowledged to be a
fair significance to attach thereto.

The assumption of the average Allopath that the
law of similia is supposed to be brought into requi-
sition to act at the bedside to the exclusion of all
other natural laws is too puerile to merit consideration.   Does it keep the patient upon his couch?
Does it keep him warm?   Does it nourish him?
Does it provide the other necessary ameliorating or
curative measures when mechanical helps are called
for, as in fractures ; or chemical helps, as in poisoning?   Neither does the Homœopathic school pretend
that the law will be universal in the sense that it
will trench upon the limits and sphere of other laws.
The law of gravitation is called a universal law, but
it will not act or produce results in the matter of
chemical combinations of bodies, but the law known
as the law of *multiple proportions*, or when one chemical is capable of forming several combinations with
another, that it always does so under the law of

definite multiple proportions; thus, neither the law of gravitation nor the law of similia has any force in the matter of the chemical combination of bodies. We may understand then that each and all laws are operative in their limits, and *exclusive in their application there,* and universal, in the sense that whereever the necessary conditions can be brought about (as the bringing the necessary chemical solutions together, properly diluted and at the suitable temperature, will establish an exhibition of the working of the law of chemical affinities,) an operation of that law will ensue.

Concerning the charge, which we have numbered "two," let us remember the assertions of Bichat, already quoted in these pages, also that of Sir Astley Cooper, who said: "The science of medicine was founded on conjecture and improved by murder." Dr. Andrew Combe, an eminent English physician, said: "In fact medicine, as often practiced by men of undoubted respectability, is made so much of a mystery and is so nearly allied, if not identified, with quackery, that it would puzzle many a rational looker on to tell which is the one and which is the other." Where are the Homœopaths who are greater offenders than these, and yet Sir Astley and the rest remained unpunished for their declarations.

Those assertions, which we have numbered "one,"
we will now entertain by dividing it into three
parts—First, What of this experience of the medical profession, which it is asserted Homœopathy
offends by being at variance with? Those who have
carefully perused the preceding pages will be apt to
opine that it would be an exceedingly difficult
matter not to be at variance with some of the many
experiences of Dogmatic medicine, but let us see
what Dr. Paris, an ex-president of the Provincial
Medical and Surgical Association, says of this experience of the Dogmatic school. He says: "We
shall not be surprised at the very imperfect state of
the Materia Medica, as far as it depends upon what
is *commonly called experience.*" He says: "Dr. Ray
attempted to enumerate the virtues of plants from
experience, and the system serves only to commem-
orate his failure; Vogel likewise professed to assign
to substances those powers which had been learned
from *accumulated experience;* and he speaks of roasted
toad as a specific for the pains of gout, and asserts
that a person may secure himself for the whole year
from angina by eating a roasted swallow." Very
possibly Homœopathy may be "at variance with it."

The statement that Homœopathy is opposed to

science is untrue, as witnessed by the fact that the recognized principles upon which science is based are acknowledged and followed by the toilers in the vineyard of Homœopathy, and, since the results of the labors of these workers are to develop new facts in Homœopathy and to advance its usefulness, as exampled in its unfolding and developing the effects and character of drugs as regards their action upon the human system, Homœopathy, therefore, claims to be strictly in accord with science, and to experience new and reinforced strength each year by the advances and discoveries of the natural sciences.

If a science is a "positive knowledge of facts based upon a law or laws in nature," then, from what we know of Homœopathy, we recognize it as a science of therapeutics, (not exact and perfect in all its departments for what science is,) but in that it represents positive knowledge, first discovered and first developed by the new school in medicine in the light of and while standing upon the law of *similia*. Therefore are those justified who assert Homœopathy to be a science.

Let us now make some inquiry as to the validity of the claim that Allopathy has to a scientific basis. Let us see, from the testimony of those the most prominent in their ranks, if the dominant school can, in the field in which it tries to assert itself, to the exclusion of its rival, show a possession of *positive knowledge.*

The eminent Moliere says, "most people die of their remedies and not of their diseases."

Dr. Pereira, in his great work on Materia Medica, says, that Dr. A. T. Thompson and Dr. J. Murray regard mercury as a stimulant, Dr. Cullen asserts it to be a sedative, and Dr. Orfila believes that it is neither.

A standard Allopathic authority, "*Taylor on Poisons,*" says, "Four grains of opium * * * * killed a man in nine hours." Pereira's Materia Medica says that the common dose of opium is from one-half a grain to five grains, but these doses "are by no means to be regarded as the limits."

These drugs—mercury and opium—we know are not new and strange ones, but, on the contrary, have been used by the dominant school for hundreds of years; that they are drugs most commonly employed is also a patent fact, and one which by turning to "Wood's Practice" (a recent and acknowledged standard Allopathic authority in America) we shall illustrate. We find that this work in the treatment of one hundred and fifty kinds of sickness directs the use of mercury in one hundred and three. We now can judge of the *positive knowledge* possessed by Allopathy, and of its position as regards science generally.

To the asseveration made by Dogmatism, and held up before the world as a heinous offence, that Homœopathy is " opposed " to " common sense," let us now give careful attention. Dr. Wm. B. Carpenter, the eminent physiologist, accepts Dr. Reid's definition of the term *common sense*, who says that it is " the first degree of wisdom," that it is to "judge of things self-evident" in contrast to the office of " ratiocination " or "second degree of reason," which is " to draw conclusions that are not self-evident from those that are." Dr. Carpenter says further: * "This is the form of common sense by which we are mainly guided in the ordinary affairs of life; but inasmuch as we no longer find its deliverance in uniform accordance, but encounter continual divergencies of judgment as to what things *are* " self-evident "— some being so to A whilst they are not so to B, and others being self-evident to B that are not so to A— *it cannot be trusted as an autocratic or infallible authority.*"

---

* Mental Physiology, p. 472.

To what extremities are the sectators of Dogmatism driven when they must conjure up to their aid such an unreliable witness! It will be seen that in this connection they completely ignore that very respectable and rapidly increasing power the clientage of the Homœopathic profession, which we find represented in the social and political economy of this country in all the higher walks of life. Such for example as a commander-in-chief of the Federal army in the last war, a chief executive, and many cabinet ministers, those the most prominent as poets, authors, ministers, lawyers, as well as others influential and prominent in commercial circles, are found employing and pronouncedly in favor, and always willing to rally in the cause of the new school in medicine.

In America, alone, Homœopathy has millions upon millions of supporters, and it is also a notorious fact concerning the Homœopathic laity of America that if this element, as compared to the number of the clientage of the Allopathic profession, is as is one to seven, that the comparative proportion of these two elements in the more enlightened and cultivated classes would be as is three to eight. Those conversant with the facts will recognize that justice is not here given to the minority, but we would prefer to underestimate their side of the question rather than the other.

This element is here referred to, for we have been giving some attention to the assertion that Homœopathy is opposed to *common sense;* now, while it is an insult to treat in this manner this very respectable and influential element, we shall see that its influence has not been wholly ignored by the Allopathic profession.

We remember the drift and tendencies of Dogmatism at the time of Hahnemann's withdrawal, and for a time thereafter. From this condition of the misdirection of the medical art, under the influence and aims of Dogmatic instinct, some competitive influence is found to steadily and unceasingly be at work upon the Dogmatic mind, an influence that ruthlessly robs their hypotheses and assumptions of their fanciful coverings, exposes them in their bare falsity and so leaves Dogmatism nonplussed and dumb. Again, by exhibiting before Dogmatism the heretofore unequalled results which this, its competitor in the medical arena, has produced, the new school imparts to it an unconscious

14

desire to pursue new directions in medical thought and research.

Although the average Allopathic mind will refute the statement just made we shall presently see the steady approach of the Dogmatic school to the positions of the new, until it has stood for some time upon its threshold, unwilling to enter and identify itself with what it has so opposed, and towards what it has displayed so much hatred and malignity. Nevertheless, adverse to longer permitting the new school to monopolize its legitimate fruits of original research and labor, and forced by the pressure which an intelligent public mind will often exert, it seizes upon and asserts what it has not displayed and does not display, any natural leaning or attachment for, acknowledging the necessity of prescribing, for the good of its own future health, the taking of this expedient, which it has, as we shall now see, swallowed, and with difficulty succeeds in retaining within its economy.

CHAPTER IX.

"OH! BEWARE, MY LORD, OF JEALOUSY; IT IS THE GREEN-EYED MONSTER WHICH DOTH MOCK THE MEAT IT FEEDS ON."

It is easy to recall what, from the inception of the new school and thereafter, were its most prominent marks of distinction—its law, similia; the necessity for the proving of drugs upon the healthy; the administration of such small doses of a drug, that the physiological or perceptible drug effect upon the healthy system by such dose was impossible, (as purging, vomiting, etc.;) the favoring of preparing doses that, from their simplicity, as well as for the reason that the method originated with the new school, were ridiculed as were those who used them; these all we recognize as the principal features which are characteristic of the new school of medicine.

Let us now quote from some of the Allopathic authorities, which, to-day, are standard works in their school. Dr. Symond's Encyclopædia of Practical Medicine, vol. IV, p. 375, says: "Upon this ground we are disposed to suggest the use of strych-

nine in tetanus; not that we have become followers
of Hahnemann but that it is a *simple and undeniable
fact* that disorders are occasionally removed by
remedies which have the power of producing similar
affections."

The well-known Allopathic work to which we
have already referred, Dr. George P. Wood's Therapeutics, vol. I, p. 32, declares: "The same medicine
may produce opposite effects in health and disease.
Thus cayenne pepper, which produces in the healthy
fauces, redness, and burning pain, acts as a sedative
in the sore throat of scarlet fever. A concentrated
solution of acetate of lead acts as an irritant, while
the same solution, very much diluted, will act as a
sedative."

That the subject of "drug-proving" upon the
healthy" is receiving attention by the Allopathic
profession, at this late day, may be shown by referring to a work upon "*Therapeutics*," by Dr.
Sydney Ringer, a work that has run through many
editions in a few years, and has, perhaps, met with
a larger sale than any other medical work in this
decade. The basis of this work is the study of the
action of "*individual drugs*" upon the healthy; and
the preface to the sixth edition tells us that the action

of the drug is studied upon the skin, mouth, intestines, blood, etc. The results of his study are remarkable, inasmuch as he not only finds, in most cases, the same remedies most efficacious in certain diseases that Hahnemann did more than fifty years ago, but, like Hahnemann, he asserts that a minute dose of this medicine acts more promptly and surely.

Dr. Ringer's treatment, then, for fever and inflammatory conditions is not, as in Hahnemann's time, bleeding to fainting, purging, and the like, but he says, "The power of aconite to control inflammation and to subdue the accompanying fever is remarkable."* He also says a solution of bichloride of mercury, "of a single grain in ten ounces of water, in doses of a teaspoonful, is *very efficient*" in diarrhœa.*

Here, then, we have it, from an authority in Allopathy whose views and methods, if we judge from the liberal patronage his book receives, are more sought for than those of any other Allopathic author, that one-eightieth of a grain of mercury is the proper dose in this disease! What intelligent Allopath will asseverate that one-eightieth of a grain of this chemical will produce a drug or physiological effect?

---

* Hand-book of Therapeutics, p. 422, op. cit., p. 236.

This is only one illustration of many, and serves the purpose, in that two of the principles for which Homœopathy has so resolutely fought are here undeniably conceded by its antagonist, viz., first, the necessity of drug proving upon the healthy, and, second, the efficiency of the minimum dose.

We will now make some excerpts from a paper presented by a prominent Allopathic physician, Dr. Henry G. Piffard, to an Allopathic medical society, "The New York Academy of Medicine," in the year 1877. He says here, in speaking of the method of preparing medicines by trituration, that, at the present time, the United States Pharmacopœia directs that but six preparations be triturated. The first four mixtures to be rubbed "together until thoroughly mixed," one mixture is to be rubbed "together into a fine powder." The most explicit direction being imposed upon the sixth, that of mercury and chalk, the direction being that they should be rubbed "together until the globules cease to be visible." He then turns to remark concerning the homœopathic triturations, and repeats the directions of the Homœopathic Pharmacopœia which the Homœopaths employ and acknowledge as their authority in pharmaceutics.

He says that, "as many of these triturations contain, in a convenient bulk, the doses that we commonly prescribe, they at once become eligible preparations for our use, provided they are as convenient to dispense, and as uniform and certain in their effects, as the preparations which we ordinarily employ."

" In order to settle in our own mind the relative value of trituration, we have, during the past year, carefully investigated a few of them, more particularly those of mercury, arsenic, and iron." " The first chosen for trituration was the 1^X trit. of merc. viv., and it was compared very naturally with our own analogous preparation, the hydrag. c. creta. Examined under the microscope, the mercury in the former was found to be in a state of extremely minute subdivision, the majority of the separate globules being smaller than red-blood corpuscles, and many of them so small as to be endowed with Brunonian movement. Fine samples of hydrag. c. creta. obtained from Broadway drug stores were then examined. These were found to vary greatly in their gross appearances, and likewise, under the microscope. In some there appeared to be a notably larger proportion of mercury than in others, and in all of them the metallic globules were very much larger in size (average) than those in the merc. viv."

*   *   *   *   *   *

" The fact of the absorption of solids at one time deemed impossible has now been so thoroughly demonstrated, and especially as regards mercury, that we are prepared to understand how minute subdivision will facilitate absorption when larger particles would pass the bowels without effect, or simply produce local effects varying with the nature of the drug employed. We should therefore expect that a given quantity of drug being used, the promptness of its specific effects would vary immensely with the size of the particles of which it is composed. If we submit this rational conclusion to the test of clinical experience in the cases of hydrag. c. creta. and merc. viv., we will find it easily verifiable. The latter given in doses containing the same amount of mercury, as the former will produce the more prompt and decided effects." He further goes on to describe the specific or bad drug effect produced by larger doses of drugs, which is obviated by using a trituration in small doses.

In speaking of another chemical salt which he has experimented with in trituration, he says, in its

common dose, that it " frequently produces gastral-
gia and diarrhœa is well known, and personally we
believe that this is due to the local irritation produced by them, and not to any elective action of the
drug. Since we have used the trit., however, in preference to the ordinary pills, patients more rarely
complain of disagreeable sensations. We have
further been enabled to materially reduce the size
of the dose in order to obtain the desired effect. In
other words a large proportion of the drug is utilized
for specific purposes, while a small amount remains
to give rise to local irritation."

He recommends not only the very large number
of chemical combinations he so administered, but
also asserts that many vegetable substances can also
be, to more advantage, administered in this preparation, both as regards *greater facility, accuracy of
dosage,* and *certainty of effect.* He also says, with
"The city practitioner at night a few triturations,
carried in a pocket-case, would often enable him to
meet emergencies without the hour or half-hour's
delay in having a prescription prepared." "To the
country practitioner they are invaluable." In conclusion, he is willing to give due credit to the
Homœopaths for their development of this mode of

preparing pharmaceutical substances, but not willing to allow them to longer enjoy a monopoly of their use.

After this very extended consideration of these several assumptions, charges and pretexts of the representative bodies of the dominant school, in reference to its rival in medicine, we will be content to rest here, after referring once more to the assumption numbered six in the last chapter.

Here, it will be seen, the reference to themselves, of the Allopathic faculty, is by the term "regular," and as we so often find this term so used by them, and that of "irregular physician," by them, applied to those of the Homœopathic profession—in fact, as these are the terms which Allopathy most frequently employs to distinguish the two schools, let us here quote the action taken upon this subject in June, 1881, by the National Homœpathic Society, ("The American Institute of Homœopathy.")

The president, (J. W. Dowling, M. D.,) in his annual address, states that "no medical body has ever given a definition of the phrase "*regular physician*," and as the members of this body claim to be regular physicians, and as we have the same right to define the words regular and irregular, as applied to medical practitioners, as any other organization, and as this is the oldest national medical organization, and as there is much in the right of priority, would it not be well for this institute, taking Webster's Unabridged Dictionary as its guide, to define, for the benefit of the medical profession at large, the phrases "regular physician" and "irregular physician?" Webster defines the word regular as, conformed to a rule; agreeable to an established rule, law, or principle; to a prescribed mode, as a regular practice of law or medicine; governed by rule or rules; *steady or uniform in course;* not subject to unexplained or irrational variation; instituted or initiated according to established forms or discipline, as a regular physician. Taking the history of medicine for the past fifty years as our guide, I would ask to which system does the term regular, accepting Webster's definition, apply?"

"*Resolved,* That the president's definition of the words *regular* and *irregular,* as applied to schools and practitioners of medicine, be adopted by this Institute as correct."

CHAPTER X.

Of the many attacks upon the new school of medicine, by the adherents of the Allopathic faith, which are constantly and unremittingly appearing, a book, which has been presented to the public very recently, under the title of "Medical Heresies," written by Professor G. C. Smythe, of Indiana, not only deserves to stand among the rest as one of the most able of them, but it also presents to the reading public a feature of novelty, inasmuch as in its preface the author informs us that he takes the very unusual course of conducting his criticism of Homœopathy fairly and from a "scientific standpoint," and "*without ridicule.*"

There is something so rare and uncommon in this announcement that we turn to a perusal of its pages with increased interest. In the beginning of the first chapter we learn that with the disappearance of the Brunonian methods occurred the "death of Dogmatism," and that from this time its successor, the "rational" school, has appeared to fall heir to its estate. For some reasons this might seem a very

221

advantageous position for the dominant school in medicine of to-day to assume, inasmuch as it in this way at one master stroke assumes no further responsibility for its most embarrassing sins of omission and commission.

However, when we look carefully at this new position asserted for his school by Professor Smythe, we fear there is but slight chance that he will be indorsed by his brother sectators of the Allopathic faculty. This school, in these latter days, has seen one after another of its strongholds assaulted by Homœopathy, until at this time it has but little (as we have seen) to rely upon and flee to, except its claim that being the school of antiquity, and, therefore, representing the "assembled wisdom of the ages," its competency in medicine is *ex-cathedra* (absolute and ultimate.)

It is unnecessary if we have followed the preceding pages here to take further space to show why Professor Smythe will not be sustained, for our reader will at once recognize how very patent is the fact that Allopathy will refuse to permit the Homœopathic school to be designated (as it would by this admission) the *elder school in medicine*.

If we were to consider seriously this statement of

Professor Smythe's, we should wish to learn further of the incidents and time of the alleged " death " of Dogmatism, just when did it occur? Surely the decease of an organism thousands of years of age would have excited some remark at the time, and yet the members of the dominant school in medicine, through the first decade of the nineteenth century, make no mention of the occurrence. Again, when " death " had fully asserted itself there would have been suspension of the natural functions, and also some exhibition of the new school (which Professor Smythe says succeeded to the Dogmatic) of its rights of succession.

At different periods the Dogmatic sectators have arrogated to their organization different names. We have seen that for hundreds of years, from about the twelfth to the fifteenth century, it preferred to be called the Galenic school, not that it was not the same Dogmatic school, for neither themselves nor Professor Smythe deny this; for the Galenic school is, and has been, recognized as possessed of the same distinctive marks, and to be the same identical Dogmatic school of a thousand years before.

Why should the Allopathic school of the nineteenth century claim to be a different school? That it has the *same distinctive marks and characteristics*, the reader knows full well. What does Professor Smythe then claim as different in ancient Dogmatism in its tendencies and bias ? Even the term applied to his new school, the " rational " school was used to designate Dogmatism two thousand years ago. However, as we have said, since there is no probability of Professor Smythe being supported in this assumpsion of his by his colleagues in the Allopathic school, there is little need of our giving further attention to it.

It will be remembered by the reader that the position assumed in these pages, as regarded the true relation of Hippocrates to the profession, was that, as the Dogmatic school was not founded until after his death, therefore, the claim that he is the true representative and paternity of Dogmatism is not tenable; and, further, it will be remembered that the effort was here made to indicate which part of the profession has the right to regard him as its paternity. Now, some, from reading here, may have thought they found good reason to believe that, by these definitions, the Dogmatic sect was without known paternity; to such we would, for their comfort, suggest that they turn to the twenty-fifth page of the work "Medical Heresies," where they will find the statement that the ancient Egyptian priest-physicians were taught "venesection by the hippopotamus, which, it is said, performed this operation upon itself by striking its leg against a sharp reed and opening a vein in this way, and, after the blood had flowed as long as it *thought proper*, filled the wound with mud," and in this connection intimate, if, from this statement and authority, it would not be well to consider if, in the hippopotamus, we have not the father of Allopathic medicine.

The professor, after devoting the first ninety-five pages to the subject of medical history down to 1790, gives over the remainder of his book of two hundred and eighteen pages to a consideration of Homœopathy, and, by this complete consecration of his work, as regards the nineteenth century, to this subject, we early distinguish the position which the new school occupies in his subject, in his opinion. The opening pages upon this subject allude to the new school as " Dogmatic," and, at intervals on through the subseqent pages, reference is made to that "exclusive dogma" or law, *similia*, which so sorely tries the soul of the average Allopath.

15

Our reader has become acquainted, from the quotations in these pages, of the Supreme Allopathic Councils of this country and Great Britain, with the chief and almost only offense charged to the new school—the basing of their practice upon an "exclusive dogma"—in asserting and acknowledging which they are found "guilty" of conduct "unbecoming and unworthy an honorable physician."

While reminding our reader of this, and while pointing to the position of Professor Smythe, just remarked upon in the last paragraph, we wish to refer to the very extended and exhaustive effort made by the Professor in the subsequent pages, where he quotes pages by the score from Homœopathic literature, to prove what?—That there are many Homœopathic physicians who use the "high attenuations," and rely upon them solely in the treatment of disease; that others use the medium and "low attenuations," and then he submits proof to establish the fact that there are Homœopaths who resort to mixed prescriptions and drugs in certain instances.

Here, the Professor says, we have, in the Homœopathic family, divisions, and the following of a variety of courses, with different members of this school

For this work, by so eminent an authority in the dominant school, and for his successful endeavor to overturn, what, above all others, is the charge brought against the new school by the old, to wit, that it is based upon an "exclusive dogma," etc., every member in the new school in medicine should acknowledge his gratitude.

The Homœopath has learned that it is of no use for him to refute this charge, for it is as often repeated thereafter, and with no more recognition of such refutation than if the traducer were deaf, but now we have this Allopathic authority, who asserts and proves beyond the shadow of a doubt to the reasoning mind, that *the new school of medicine is not "exclusive."* It has a law and principles upon which it stands, but by the evidence here adduced by Professor Smythe we find it conclusively established that there is " liberty of opinion and action " in the new school of medicine, and no tenet or supreme council to gainsay it.

Our reader may now, after turning to the exhaustive proofs upon this point, adduced by the Professor, inquiringly add, what is there now left among those objections and charges brought against the new school by Allopathy, that has not, in one way or another, met with complete refutation ?

If the reader will turn to pages 101–102 of "Medical Heresies," he will find the words: "It has been the rule with all schools of medicine, and will continue to be so for all time, that the therapeutical appreciation of remedies is based upon the pathological conditions known to be *or supposed to be* present in the case to be treated." While it is unfortunately true, as the Professor acknowledges to be the case, that disease is, and always has been, treated by the Dogmatic school according to their theoretical notions of the character "*supposed to be present*" of the disease, the new school, as we have already seen, is established upon better and surer methods of reaching and overcoming disease.

However, we find this point waived here, and we see the professor following the plan so often pursued, when it is held up to the Allopathic physician that the law of *similia*, and the other incidental and useful principles of Homœopathy, are now acknowledged and indorsed by the most eminent of the Allopathic school, of showing his dislike for the school which these principles represent, leaving the questions and devoting himself to a consideration of the views and theory of Hahnemann concerning the method of action of the Homœopathic remedies,

denouncing them and their originator, and following upon this with a general denunciation of the
new school of medicine.

We, therefore, find the professor proceeding from
this page (102) on to make liberal quotations from
Hahnemann's writings, and to labor by these to
establish what has been, and is already, recognized,
viz., that Hahnemann endeavored to explain, by a
theory of his, the mode of action of the Homœopathic remedies. Hahnemann does this by supposing, with *Stahl*, that the soul is competent, by its
influence upon the physical body, to produce abnormal or diseased conditions. We shall recognize
the striking similarity of his views to those of *Stahl*
if we quote these words of Hahnemann's, from the
thirty-fourth and thirty-fifth pages of the introduction
to the Organon, he says: "Diseases will not * * *
cease to be dynamic aberrations, which our spiritual
existence undergoes in its mode of feeling and acting; that is to say, immaterial changes in the state
of health." This is the basis of the "transcendental
pathology" which Professor Smythe arraigns, and
for the production of which poor Hahnemann is so
thoroughly condemned.

Of these theories of Hahnemann's we may say

that they were consistent with the most advanced facts and discoveries of his time, but no position or division of the new school of to-day, accepts his theories, but alludes to them only as evidence of the ability of their originator, and though they admit them to be erroneous in the light of modern science, yet they ask how it could have been possible for Hahnemann to take the facts of science which were to be brought forth after his death into consideration in his views of the causes of disease.

However, let us here point to one marked difference between these views and theories of Hahnemann, and those of the sectators of Dogmatism. While with the latter, the diseased conditions "supposed to be present" were a basis of treatment, with Hahnemann the treatment was to be strictly and always in accordance with Homœopathic principles and the law of similia. In other words, these views were given to us to explain for our delectation the manner of the action of remedies.

We shall soon see, as we have said, that the different positions of the Homœopathic school do not accept or defend these theories—not that there may not be some Homœopathic physician who does accept them, for he is at liberty to do so if he so chooses—

but speaking for the Homœopathic school of medicine of this time, we say with confidence that the obsolete views of science of fifty years ago are not asserted by the new school, and any insinuation that this school does not live in the light of and keep pace with the times will be emphatically denied by its adherents.

This applies to the views and theories of the past, but not to those truths and facts which have been from time to time brought forward by their discoverers and left as a legacy to the future generation of man, for these are immutable, and to endure as they always have, and to eternity.

Because of the entertainment of these theories, then, poor Hahnemann is to receive the condemnation of futurity, while Stahl, whose theory of the cause of disease was in no essential much different, is still accorded a high place in the history of the Dogmatic School of Medicine. When we come to consider that Lord Bacon "gravely prescribes henbane, hemlock, mandrake, moonshade, tobacco, opium, and other soporiferous medicines as the best ingredients for a witch's ointment," and that Sir Walter Raleigh firmly believed that "the old crone, with her tall hat, crutch stick, and black cat nestling

on her shoulder, was one who had dealings with
the devil, and who, through the might of Satanic
aid, could scatter the seeds of misery broadcast
wherever she listed;"* and when we know that these
are prominent and venerated men of science of the
past, and that no such absurdities can be laid at
Hahnemann's door, it seems a little strange, in view
of the immense good Hahnemann *did do* in his
unfolding of the law of similia in its application to
diseased conditions, and of his other improvements
in the healing art, which the Allopathic authorities
now acknowledge; that even a greater position in
letters and in science cannot be accorded him by
them than is given to the average worker in the
field of science, during the morning of the nineteenth century.

Farther on, at page 127, Professor Smythe, in
combatting in the most valorous manner these
theories of "transcendental pathology," says: "If
our organs are not in good working order and our
circulating fluids are depraved, our *vital force* will
be *lowered,* but *will not be and cannot be diseased,*
Hahnemann and his followers to the contrary notwithstanding."

* Ewald.

We feel exceedingly sorry to see the Professor submit these views, and chiefly for two reasons. The first, because they are false, and as he is writing in the "middle of the afternoon of the nineteenth century," there is onus attached to such falsity that will not apply to the case of Hahnemann. Secondly, we are sorry, because they are not only false but unfortunately are unmistakably a part of the Brunonian theories of sthenic and asthenic conditions of the vital power. The eminent Dr. Alison, who died in 1859, also asserted, in one of his very last works, that inflammation was "a local increase of a vital property," and treated this by bloodletting to lower this vitality. All this is as plainly Brunonism as anything can be, and now after the Professor has, with much solemnity, read the funeral sermon over the grave of *what he thought* was Brunonism, we here find it residing within his corporosity, apparently as lively and robust as ever.

The professor should know that already there are many in his school of medicine that have given over the idea of the increase and lowering of the vitality as being its only possible variation. At the present time very different ideas of the changes of the vital power are gaining ground, and before many years we shall see the Allopathic profession, regarding the vital power as undergoing aberrations, aside from such rises and falls as increased or decreased strength may induce.

The next chapter is occupied with a consideration of the provings of remedies. These he is not inclined to think well of, and chiefly because there can be no " therapeutic force " in the Homœopathic preparation of medicines. On page 150–152 he gives us his computations as regards the Homœopathic methods of preparing dilutions and triturations. The first difficulty that he experiences in this is in the preparing of the thirtieth dilution from the mother tincture, he tells us that " it is unnecessary to pursue the steps of this calculation," but that to effect the results will require millions upon millions of vials, and this he attempts to represent by the figure one followed by sixty-one ciphers; nor is this half the difficulty, for there must be, he says, twelve times as many strokes as vials, and would take a man who works ten hours a day, seven days a week, and makes sixty strokes a minute, " 661 quadrillions, 882 trillions, 919 billions, 336 millions, and 1,050 decillions of years " to prepare this. Also, the professor tells us that to manufacture what would be contained in the above-mentioned vials would be required of alcohol alone 10 quintillions, 774 quadrillions, 186 trillions, 46 billions, 511 millions, and some thousand and more hogsheads full. Now I wish to answer these statements and objections of the professor's satisfactorily, and to do so will take this way.

The City of Washington has poor hospital facilities. While in New York city the hospitals have one bed to every 117 inhabitants, those of Philadelphia one bed to every 250 inhabitants, those of Baltimore one to 150, Washington has only one hospital bed to every 2,000 inhabitants. The homœopaths recognize this fact, and are striving to increase the number of hospital beds here by engaging their best efforts to establish a Homœopathic hospital.

Now, if the professor will agree to make half of a purse of five thousand dollars, the author will pledge the other half, providing we further agree that if the author cannot, according to the rules of Hahnemann, prepare the thirtieth dilution from a mother tincture and alcohol, so that the medicine, alcohol, and all the paraphernalia used in the entire process shall be completely encompassed in a pint pot, and that the whole process of this production of the thirtieth dilution, after this manner, will not take more than two hours; if the author cannot accomplish this under these conditions, then the purse of five thousand dollars shall go to increase the beds in such Allopathic hospital in Washington as the professor directs; while if he succeeds in accomplishing what is above indicated, the five thousand dollars shall be placed to the credit of the fund for the new Homœopathic hospital that is to be. An early acceptance of this offer will be gratefully acknowledged.

The conclusion of Professor Smythe that the attenuated doses of Homœopathy, from their minuteness of subdivision, cannot produce effect is an illustration of *à priori* reasoning which he, as an exponent of Allopathy, is consistent in making, but which all Homœopaths will thoroughly repudiate. Please tell us, Professor Smythe, is the contagious principle of scarlet fever any more gross than the highest potencies of Homœopathic medicines? What Allopathic authority has been able to show the size of malarial poison, or to isolate and demonstrate it? What can you say of the poison with which the mad dog inoculates his innocent victim? Because it exists in so minute and fine a condition that microscopical and chemical tests will not demonstrate it, will you also say that it is too minute a substance to be possessed of power?

The eminent savant, M. Davaine, submits to the French Academy these results of experiments by himself. By the injection of septicaemic blood, he says: "The blood of one rabbit killed by the *one ten millionth part of a drop* was injected into five rabbits, in doses of *one hundredth millionth, one billionth, one ten billionth, and one hundredth billionth, and the trillionth of a drop—all died* within twenty-five hours."

.

CHAPTER XI.

" But man, proud man,
Drest in a little brief authority,
Most ignorant of what he's most assur'd,—
His glassy essence,—   *   *   *
Plays such fantastic tricks before high Heaven,
As make the angels weep."

The next (last) two chapters are especially devoted
to the making clear, with the help of abundant quotations, (the current Homœopathic literature being
called upon to furnish pages by the score,) of the fact
that there are open questions in the new school of
medicine upon which the Homœopathic profession
do not unite, but on the contrary make these unsettled questions subjects of controversy. These
"diverse opinions," "obstacles," and "follies," as
Professor Smythe denominates them, are discussed
by the Professor in a manner to which the average
Homœopath must necessarily take exception. The
reasons for the misapprehension of the Professor, in
his discussion of the principles and views of the
Homœopathic school, are chiefly two; and do not
surprise us when we consider that they would
naturally influence and bias any adherent of the

(239)

dominant school in his examination of a medical
system.

We do not wonder that the Professor misapplies
the fact that there are so many views advanced upon
the subject of what shall be the proper potencies
and number of repetitions of the dose with Homœopaths, but the Professor should remember that he is
now investigating matters without the pale of his
school, and that the Homœopathic school of medicine presents new features to him, inasmuch as its
adherents are at liberty to entertain most extraordinary views upon these open questions without in
any way affecting their standing or fellowship with
the other members of the profession. This is illustrated, even in the views of some Homœopathic
physicians in regard to the mode of action of
Homœopathic remedies; for these physicians, (being
exceedingly few in number,) having a prepossession
for the religious sect of spiritualists, incline to believe that the action of Homœopathic remedies is
consistent with their religious faith.

The very opposed views of prominent Homœopathic physicians upon the question of " potency,"
and the relative value of the " key notes," are other
instances of this liberty of action, for we now see

that, unlike the Dogmatic school, the new school is so broad and comprehensive in its genius that though the most opposite views obtain with different of its adherents upon these controversial matters, and though each view is defended as regards the opposing ones with much warmth and vigor, none of these controversalists are, in any sense, the less orthodox in their relation to the new school of medicine than any other of the followers of Hahnemann.

In fact, we only acknowledge by our position in these matters that we are sure that in this direction much is to be learned. Here our science is yet imperfect, and these are the methods of labor, each in its own way, for the further development of this branch of the system of Homœopathy, and as from the results of such scrutiny, investigation, and debate in the Homœopathic ranks, upon these open questions much has been developed, therefore, confidence has been and is felt that more value and profit will result from the continuance of these efforts.

The Professor will therefore see, that by his exhaustive efforts in this direction, he has raised to himself a man of straw, which, with evident satisfaction, he most thoroughly demolishes, for we can-

16

not but see that his bringing forward some statement advanced by a Homœopathic brother, with
considerable buoyant confidence, does not, therefore,
establish it as the true and accepted statement of the
Homœopathic school upon this subject, to be placed
beside the principles and law of Homœopathy which
Hahnemann asserted, and which the adherents of
this school so zealously uphold and vindicate; on the
contrary, he must take these statements as do the
profession of the new school—to be tried by the
ordeal of examination and experience, and then be
accepted for what they are worth.

The Professor tells us that these differences give
us two species of Homœopaths, the Eclectic-Homœopaths and the Straight-Jackets. As it is just as well
to stand upon the facts with this as well as with all
other matters, let us investigate this question, that
we may judge more intelligently of the subject in
question.

It has already been remarked that, in or about
the year 1830, there appeared marked differences as
between Hahnemann and those his most prominent
colleagues. From what we have learned of the
course of the dominant school towards Hahnemann,
and of the unrelenting manner with which he was

hunted down by them and persecuted, when we
come to consider that, in the earliest days of the new
system of medicine, that for years, in his isolation,
he was the tangible embodiment and representative
of the new school, it is evident that such experience
as was his could not but embitter, roughen, and
materially transmute many of those lovely and
prominent traits of his natural character and disposition.

Therefore, when, in following carefully the history
of Homœopathy, we find that in later years, and
particularly subsequent to the year 1830, at which
time he has reached his seventy-fifth year, he does
not seem to be inclined to tolerate friendly criticism
from his confréres, and even presents many of his
ideas to the then rapidly growing profession of his
school in a dictatorial spirit, and seems to feel it a
matter of the nature of a personal insult that any
part of his views and observations are not received
unquestioned, we do not acknowledge these as
evidences of the true genius of Hahnemann, but as
the combined result of his accumulated years, and
the imbittered and petulant mental status superinduced from the years of subjection to Dogmatic
malevolence.

The differences in the profession of the new school began to take form previous to this time, when, with Hahnemann at Cœthen and Dr. Moritz Mueller, occupying the position vacated by Hahnemann, at Leipzic, the Homœopathic journal, the "*Archives*," etc., nominally under the editorship of Drs. Stapf and Gross, but, in fact, managed by Mueller, was attacked by Hahnemann. In these later years of Hahnemann's life work, we find him constantly preferring the higher preparations of the Homœopathic remedies, and, after a time, we find him using the thirtieth potencies almost exclusively. When spending the evening of his days at Paris he declared himself in favor of these higher preparations, but we find that the Leipzic Homœopaths do not follow him, to the extent that they insist upon the exclusive use of the higher preparations.

We find, then, this difference in the Homœopathic ranks, that the quick and active mind of Hahnemann, so alert and prompt to seize upon the legitimate and consistent outcome of his initiative, had, in declaring in favor of the higher preparations of medicines, induced a number of his followers to accompany him upon this new ground of his choosing. On the other hand, Mueller and the rest, who were gathered in the more conservative fold, men of pronounced ability and talents, represented the sober, cautious, plodding, straight-forward, and unimpeachable mind in the Homœopathic school. As, down to this day even, these divisions in the ranks of Homœopathy have continued to exist, let us consider carefully the prominent issues prevailing with each portion of the new school. The term Eclectics or Eclectic-Homœopaths we shall presently see to be a misnomer as applied to the Leipzic Homœopaths. This division seemed to be more partial to the use of the lower attenuations, and to give especial prominence to pathology and the *causa occasionalis.* Dr. Frantz Hartman, a colleague of Mueller's, speaks, in his able work upon "Therapeutics of Acute and Chronic Diseases," etc., of the necessity of keeping informed of the recent improvements in diagnosis through the advancement constantly made in physiology and pathology, that the various diseases may be better understood, that their true *similimum* may be applied and the proper external conditions understandingly employed. He also says that he wishes to silence "our opponents," the Allopaths, by showing that it is imperative, if the true Homœopathic remedy be chosen, that the *causa occasionalis* and pathological state be *known*, not supposed; for, he says, a remedy seeming to be the true *similimum* might appear not to be when these elements are admitted into the case as factors; thus, a rheumatic cold, with the usual symptoms, would point to several remedies, if such cold were superinduced from drenching the skin in a cold storm, there would be a reason for the exhibition of the remedy rhus tox. Again, "what physician would not give cocculus against a febrile state, characterized by flushed cheeks and nightly sleeplessness, if he knew that homesickness was the exciting cause?" "Physical and mental weakness, resulting from blood letting, hemorrhage, waking, night sweats, onanism, venereal excesses, etc., finds a specific in china, provided the weakness is the principal suffering, and not a mere symptom of a more general and deep-seated disease."

The insinuation that these, the "low potency," or "liberal" wing, were any more favorably inclined to Allopathy than the other portion of the Homœopathic profession will be readily appreciated when we consider the treatment of one Dr. Fickel, who managed to ingratiate himself with the inspectors of the Homœopathic hospital at Liepzic, and received the appointment there as physician to this hospital. When it was shown that he had practiced imposition upon the Homœopaths, that he had no sympathy with the Homœopathic principles, and was in fact practicing bad Allopathy, he was deposed from his hospital position, and was glad to take himself away from the new school and its profession.

Again, the assertion so often made by the Allopathic profession and repeated in "Medical Heresies," by Professor Smythe, to the effect that the wing of the Homœopathic profession, which he terms the "straight-jackets," or the "Hahnemannian's," regard the views and opinions, as well as the facts and principles which Hahnemann brought to light, all as necessarily true, because he uttered them, that the "Organon" is their "Bible," and Hahnemann's utterances be absolutely, finally, and fully accepted, because of their authority, are statements without foundation and without fact.

For, though some individual physician of to-day may so incline to believe, or if there were some relatives or admirers so ardent as to be willing to stand upon this definition of what they accept relating to Homœopathy and Hahnemann, though they would still be regarded as good Homœopaths as any of their colleagues, the assumption that any considerable number of either wing of the Homœopathic profession were consonant with this is not true.

The Professor endeavors to establish this by quoting the resolutions passed by the International Hahnemannian Association. A perusal of these resolutions will develop the fact that these Hahnemannians assert here only an indorsement of the use of the high preparations and a plain and full affirmation of the *principles* of Homœopathy as enunciated by Hahnemann. The next words of Professor Smythe, after finishing the quotation, are, "The foundation of this Association and the adoption of this platform of *principles* is a return to the pure, inflexible Dogmatic Homœopathy of Hahnemann."

Although the Professor here acknowledges that it is the principles of Homœopathy which the Hahnemannians accept, carrying them in the Professor's opinion to the pure Homœopathy of Hahnemann, yet we find him in common with all other Allopaths, imposing upon this very respectable and talented coterie of gentlemen, the theories and views of past generations.

That this is a slander easily refuted we will demonstrate by taking the statement of Constantine Hering, the pupil and colleague of Hahnemann, who came to America to extend and increase the fame of the new school; whose memory stands to-day, at once, a monument to Homœopathy as well as to that wing of the school known as Hahnemannians, for none were more prominent in either than he. In the fourth American edition of Hahnemann's Organon, in the introduction by Dr. Hering, he says, that " all Homœopathic physicians are united under the banner of the great law of cure of *similia similibus curantur*, however they may differ in regard to the theoretical explanation of that law or the extent to which it may be applied," thus establishing that even Hering acknowledged the difference between accepting the principles of Homœopathy, (for it is this that makes one a Homœopath,) and the theoretical creations that were often proffered to keep these company.

The application of the term eclectic-homœopath, or eclectic, as applied to the other wing, we have before referred to as being a misnomer. To better understand the true significance of the term eclectic, as used in this day, let us in retrospective remind our readers of eclectics of ancient times. It will be remembered they were a set of physicians who could not identify themselves with any fixed principles, and, in fact, had no basis or foundation upon which to stand; in other words, there was no unition of these eclectics upon any principle, except that they asserted that no medical sect represented all the truth in medical science, and, therefore, each eclectic was to choose, from every direction, those things " which he considered best."

Beside the Eclectic, the Empiric and other ancient schools admitted that new medical truths were discovered and developed outside of their several limits, so we see that the first proposition was not peculiar to ancient Eclecticism. The pretension of these (Eclectic) physicians held that it is safer to rest it with the individual experience, reason and judgment to decide concerning the medical theories and methods of their ancient time, in fact, it was " individualism erected into a dogma." The medical sect, which in modern times, appropriated for its own use the appellation of this ancient term as applied to medicine, we shall find to possess distinguishing traits entirely different from those of the ancient sect of Eclectics.

Some years ago, there practiced in Boston, Massachusetts, a physician, Dr. Thompson, who took objection to the practice of the Allopathic physicians in the use of such chemicals as mercury and antimony, and proclaimed that in their place he used preparations of roots and herbs. He speedily acquired a decided local reputation, and other physicians accepted his position in antagonism to the accepted tenets of Allopathy.

The legitimate successor of Thompsonianism is the modern school of Eclecticism, and its principles may be stated by taking Dr. Scudder, of Cincinnati, a physician than whom none have been so prominent in the modern Eclectic school. In the introduction to his Eclectic Practice of Medicine, he tells us that " disease, wherever met, and in whatever form," is an impaired vitality, and, therefore, we not only " reject bleeding, but the use of mercurials, of antimony, etc." Dr. Scudder also says that other schools claim to have exercised, and to still exert their right of appropriating that which is best from every source, " and in this respect are truly eclectic ; " therefore, as this is not a distinguishing feature of the modern Eclectic school, what has been enumerated above should be looked to as what eminently does distinguish the modern school of Eclecticism as such.

That the term of Eclecticism has no application with the members of the Homœopathic profession is patent enough, for none of its profession are willing to recognize and assert the principles as set forth by Professor Scudder. Eclectic Homœopath has an application, however, as regards the very considerable number of physicians of the Eclectic school of medicine, who having examined the principles of Homœopathy, recognize their value, and to a greater or less extent employ them. The term "Eclectic," or Eclectic Homœopath, as applied to "liberals," or low-potency practitioners, is therefore, as has been contended in these pages, a misnomer, but applicable to another class of physicians. The assertions which attempt to affix to the Hahnemannians the full acceptance of theories harmonious with the scientific status of fifty years ago, because they are found in the writings of Hahnemann and applied to his new discoveries, are also, as we have seen, not in any sense true; these are not the distinguishing marks of these wings of the Homœopathic school, but the fact of the existence of such divisions are but admission that the developments of the natural sciences and the advancement of Homœopathy have not yet reached a stage where the profession can receive light enough to fully and finally explain what, as yet, from the ignorance of mankind, necessarily continue to remain *open questions.*

An illustration of this is the differences as regards the extent of the applicability of the law of similia, as Dr. Hering declares,* but we have every reason

---

* For instance, there are no species of burns which are more dreaded by the physician, as offering at once, by the complete and penetrating destructivity of the tissues in contact as well as by the slow and most unsatisfactory process of repair and recovery that they present, as is exhibited when accidents occur with the chemical agent sulphuric acid.

Now neither wing of the Homœopathic school of medicine would condemn the prompt external and local application of calcined magnesia made into a paste with water, for we know that such application will, by the operation of the law of chemical affinity, produce between these two substances (the magnesia and the acid) a new substance—the sulphate of magnesia, a harmless substance, and grateful change from that of the terrible irritant poison, sulphuric acid.

The prompt use of this local antidote has not only worked marvelous results in completely removing the poison in such accidents, but through the subtle and complete efficiency of the law of chemical affinities the acid is found to be so effectually reached and overcome in the minute pores and tissues adjacent to the burn that the charred surface immediately presents and unceasingly follows the most rapid processes of complete repair and recovery.

On the other hand, were a congestive disease of the inner ear to produce a neuralgia of the nerves in the face, the recognized treatment by the old school of liniments and flying blisters applied

to believe that these differences will finally inure to the benefit of this obscure question, in that aid from the controversies subsequent upon these differences will presently be manifest.

We therefore see that these differences, extant in the new school of medicine, are not disintegrating in their nature as regards this school, also the party feeling is not truthfully pictured by its opponents; on the contrary, the relations of these two factions are after this fashion. The liberal Homœopaths recognize the consistency and thorough honesty of the exclusive high potency brother who will not, under any circumstances, give potencies other than the high, sending such patients as require other treatment to some brother physician, as the physician sends his surgical cases to a surgeon, and they also notice the remarkable adeptness and skill acquired by these Homœopaths in selecting the true Homœopathic remedy.

locally to the seat of pain would not be indorsed by the Homœopathic physician, be he high or low potency in his tendencies, but, recognizing the exciting cause and its character, he would administer internally a remedy whose homœopathic action was similar to the action (both as to location and character) of the disease, and would see its prompt efficacy manifested in conquering the existing complaint.

The chidings extended to him by such, the liberal gracefully and pleasantly receives as from one consistent and honest, but when he recognizes, "ye mongrel physician," one who is ever ready to expound upon the necessity of "elevating" the practice of Homœopathy, and to improve every opportunity for sermonizing concerning the efficacy of the high potencies, when it is perfectly well known that such physician is unfortunate as concerns his natural abilities for the healing art, and the result of his work an often exhibited failure in pursuing the correct methods of prescribing, till tacitly acknowledging his lack of results he flees to the lowest potencies, and there, at last, failing completely, another and more able physician is called, who, with the simple and correct prescription, achieves prompt and immediate relief, this "mongrel," choosing to assume that his successor had not exhibited as faithful a regard for the principles of Homœopathy as did himself, by his pratings continually strives to distract attention from his own incapacity and to make such attacks as this upon his brother's orthodoxy *an apology for his own incompetence;* before such it is true enough that the liberal Homœopath poses in neither a complaisant or charitable attitude.

The closing remark of Professor Smythe, in his effort to demolish not " Medical Heresies " but Homœopathy, is: " This kind of Homœopathy will not stand the test of recent advances in science," which is further qualified on page 132, where he says, " yet its (hom.) fate is certainly sealed, and, in a few short years it will be numbered with the delusions of the past," is neither an uncommon nor unique statement, but, withal, rather singular, if we consider that, in the first days of Homœopathy, the positive assertion was often offered to the suffering Allopathic profession in Hahnemann's immediate locality, that the days of Homœopathy were numbered.

Again, when Dr. Quin defied the examining board of the Royal College of Physicians, they said, let Homœopathy alone and it will die in a year or two. Professor O. W. Holmes also said of Homœopathy, more than a quarter of a century ago, that it would very soon be a thing of the past, and now, in the year of our Lord eighteen hundred and eighty-one, the science of Homœopathy, protected and fostered by the organization known as the Homœopathic School of Medicine, exhibits in the United States twenty-three State medical societies, ninety-two local societies, seven medical clubs, thirty-eight hospitals, (containing 1,682 beds,) twenty-nine dispensaries, (treating annually 117,000 patients without charge,) eleven colleges, sixteen medical journals, and a medical profession of seven thousand physicians, in round numbers; this is the evidence upon which Professor Smythe asserts that Homœopathy's "fate is sealed, and, in a few short years," will be *non est.*

This statement is on a par with those so frequently heard and read in Washington lately, and wherever the establishment of a Homœopathic hospital seems to be among the probabilities of the near future.

We are told that no reason obtains why the Homœopaths should possess hospital facilities that will not equally well apply to the hydropaths, clairvoyants, electricians, "rubbing doctors," etc., when as regards such means of treatment as hydropathy, electricity, and the like, the facts relating thereto show these to be neither a system nor a school of medicine, but a means resorted to only in the treatment for the cure or alleviation of certain diseases or conditions; whereas, we have seen that both Homœopathy and Allopathy present valid claims as schools of medicine, inasmuch as they possess colleges where elementary instruction in anatomy, physiology, pathology, chemistry, etc.,

17

is given, the systematic and thorough education in medicine generally is prosecuted, and where the relation of the natural sciences as aids in the reinforcement of these systems of medicine in the treatment of all classes and features of disease is held up.

Therefore, Homœopathy is competent to be trusted (through its representatives in the profession) in the care and responsibility of all kinds and classes of disease, and particularly in this matter. So when we remember that by the great and critical public in no system or school of medicine is absolute authority rested, but on the contrary, every and all methods and systems are always before this potent element in the sense of being on trial; when in the light of this showing, Homœopathy lays claim to be distinguished from such restricted means of relief as hydropathy or electricity, and to resemble Allopathy, in that as we have seen, it presents a valid title as a system and school of medicine, we find every follower of Hahnemann insisting that its place in medicine is not where its adversaries would locate it.

CHAPTER XII.

Concerning the query as to the possibilities of Allopathy in its relations to the cause of suffering humanity, it would seem that it now only remains to show that no change in heart of this school of medicine, has to any appreciable extent, manifested itself within the last decade.

In undertaking to demonstrate that at the present moment the Dogmatic theories still hold the same place in the scheme of Allopathic medicine as they were found to exert hundreds of years ago, we shall not deem it necessary to give extended examples, but by a well taken illustration to establish the fact of the present position of Allopathy as regards this its heretofore prominent characteristic feature.

It is with diffidence that we now attempt to pass judgment upon the ruling theories of the dominant school in medicine of to-day, for the time is not yet at hand when they may be exposed in the repletion of their iniquity of blunder and falsity; however,

(259)

one case in point, which we shall here cite, will indicate their intended character and reach, as well as
the necessity of our manifesting doubt as to their
assumed value in their intended scope and application.

Our quotation from Professor Smythe, upon the
nature of the "vital force," which appears on page
127 of "*Medical Heresies*," where he says, emphatically, that a depraved condition of the fluids of the
body or other arising conditions represents only a
lowering or raising of the vital force, and that other
alteration does not appear with this vital force, gives
us a good illustration of what may be procured
from almost every Allopathic authority upon the
subject of *vital power*, and which is, as these authorities so often asseverate, a repetition or indorsement
of the Brunonian theory of vital force.

This doctrine of the vital movement of the
Brunonian system of medicine, from Professor
Glisson's assistance, asserted that in every muscle
there resided an especial and specific vital property,
and that contraction of the muscle was but an
arousing of this vital power.   Dr. Brown, discussing
the question of a separation of this question into
one of irritability and sensibility, (as Dr. Haller

endeavored to do,) waives this point, by asserting that the "capacity for action," residing in the muscle, was aroused by "exciting" or stirring up this vital property, and that everything producing vital action must therefore act as a stimulus or excitant. "This doctrine of vital motion," says Dr. Radcliffe, (1876,) an eminent English authority, "is, with little change, the doctrine at present in favour."

The Homœopathic scientist feels at liberty to oppose this theory, the application of which, in the treatment of disease, we have seen do such untold and irreparable injury, and the facts against this theory will most conclusively and completely establish its utter fallacy.

For instance, concerning the condition of an animal at the time of birth; your attention is directed to the fact that this animal is an ego made up of many different tissues, organs, and constituent parts, each of which, with the progress of development, soon assumes a perfectness of form as regards its intended service to the general economy, and which, at this time, has now become a typical one of its parent and race. Your attention has also been called to the continued appropriation of air and food by this animal, and regarding its growth

and development you may, by referring to the processes in the stomach, by which the food is reduced
to a state of absorption by the action of chemical
processes, capable of imitation in the chemist's
laboratory, explain to a certain extent the phenomena of this growth and development. Also its
endowments of movement by the action of the
muscles upon their levers, you may account for upon
mechanical principles and laws, and even when it is
demonstrated that certain exigencies call forth the
aid of a reserve force or power held in abeyance,
but employed when required, you may think that
this power may be but the locking up of opposing
forces within the economy by their equilibrium or
mutual balancive power.

However, conceding that these observations be
true as far as they may go, yet we shall find that
much that is remarked concerning the phenomena,
which this animal in his development will present,
will remain unexplained by the operation of chemical or mechanical laws. Following our observations
we shall see that the time presently arrives when the
growth of the animal seems to have ceased. It has
now become typical of its race and parent, and for
a certain time seems to remain in *statu quo*, appar-

ently only engaged in repairing its waste of wear
and tear of tissue until another stage presents itself,
and what appears to be a deficit account with the
external conditions is opened, an account that can
only be kept balanced by payment at the expense
of latent power and disintegration of tissue. The
organism becomes, in this way, " by the ravages of
time," daily more and more exhausted, until at
length barring the appearance of disease previously
or at this time, the animal calmly and quietly
rounds his earthly experience " with a sleep."

Recognizing this as a typical illustration, and that
the tissues of this animal in death will yield the
same chemical elements as in life, also, that the
history of this living animal has been one of *inces-
sant change* towards an end by some agency, evidently a predetermined one—for has not his history
been that of the experience and end of those of his
race (exclusive of disease and accident exemplifica-
tions) that have gone before him—we see in this
study of ours the existence of something more than
those influences which we have already recognized.
What this is in its nature and character we are
not ready to say, but we recognize it as a *directing
agency*.

How much more relation or scope this *directing
agency* or *vital force* enjoys we decline to theorize
concerning but, with these facts before us, we feel
inclined to compare this *vital stimulus* with its
powers and endowments, as regards the human
organism, to the master builder in his relations
towards his workmen, their helpers, tools, and materials. If, in delineating upon the Brunonian
theory, we have seen muscular and other actions
occur, which we are sure are illustrations of the
operations of natural laws, recognizing the truth of
the aphorism that the amount of terrestrial material
and force is immutable and unalterable, and, therefore, disagreeing with those who consider such
actions of the economy as the inherent vital power
or force, we characterize such phenomena as the
results or acts of the vital power, which are to be
compared to this vital power as is the labor performed by the workmen under his *directing* agency
and guidance to be compared to the master builder.

" Where a step stirs a thousand threads,
    The shuttles shoot from side to side,
    The fibres flow unseen,
    And one shock strikes a thousand combinations."

We might carry this illustration yet further, by

supposing that the master builder, through temporary deafness, is dissatisfied with the strength of the
blows delivered by the workmen in their labors, and
so standing by, directs their more rapid and forcible
delivery. Now, if in convoking external help to
restore the proper condition of affairs, the agent
employed the Allopathic methods, the force of the
blows would be directed against by bleeding or purging each workman that his strength might become
reduced, and so the blows lessened in rapidity and
force, while if the principles of the new school of
medicine were resorted to the deafness in the case of
the master builder would be aimed at, this being
removed, reaction and complete recovery of all the
deranged forces would immediately supervene.

Now let us regard this Brunonian theory of vital
force which, as we have seen, looks upon this property of vitality as residing in the muscle, becoming
awakened, when excited, by some stimulus, and so
manifesting itself by contracting this muscle ; let us
see if this theory is consonant with recent observations and established facts.

If according to the tenets of the Brunonian theory
there is any complaint that will demonstrate an exalted vital action, such a one as epileptic convulsions

will do so, with its rapid succession of most vigorous and powerful contractions of a greater part of the muscular system of the body. Observations have demonstrated that a division of an artery (temporal artery,) when such convulsions were at their height, allowed black venous blood to escape from it in the usual way.*

We all know that an exaltation of the vital action we connect with the sending of bright red arterial blood to the nerve centres, but black does not, under such state of things obtain, except there be sudden asphyxiation.

Again, sometime after death, and generally within twenty-four hours, the muscles of the body assume a condition of unyielding contraction which does not disappear until putrefaction is here established, this is called a condition of " *rigor mortis.*"

In the poisoning of a rabbit with strychnia, sub-cutaneously, the cramps, superinduced at the outset, developed into spasms, which carried the animal upon his toes, bending him backward against some object. After the death of the rabbit it was noticed that no change of position had occurred, the animal continuing so poised, and in a condition of " *rigor*

---

* Radcliffe.

mortis," until at the fourth day putrefaction asserted itself.*

Of course it is in place here to ask, if the Brunonian theory that the muscular contraction is representative of the vital property be true, why such contraction was not dissipated with the extinction of life, which in point of fact was not the case, for the rabbit was not displaced from his peculiar position, in which he could continue to poise only by the contraction which placed him there.

This rather extended treatment of the Brunonian theory of vital power was undertaken that the intelligent reader might appreciate how very misleading and false are its positions and assumptions. The reader would have acquired an even better conception of the prevailing theories, which to-day represent the accepted doctrines of Dogmatism in their abundance of absurdity and paradox, if some of the other theories had been set forth in the place of the foregoing illustration.

The instance here cited not only demonstrates that the most prominent characteristic of Dogmatism, the invoking of the imaginative faculty for the production of fanciful hypotheses with which to

---

* Op. cit.

enlist the prejudices of the time still obtains, but that this case in point is one that was already the accepted doctrine of the Dogmatic school prior to the nineteenth century, and has so continued to hold this position in Allopathic medicine down to this time, when it is found to be "the doctrine at present in favour." (Radcliffe.)

That it may be evident whether there be at the present day the same course of intolerance as in the past exhibited by the exponents of the dominant school towards the rival school in medicine, the latest bearing of the Allopathic faculty toward the Homœopathic profession will now be noticed.

In 1826 a part of the territory of Michigan was set apart in the interest of a university then contemplated. The Legislature of the State of Michigan, in 1837, established this university, and, in 1850, a medical department was added to it. The State constitution vests it with eight regents, elected by the popular vote of the State, to govern this university. We have already referred to the power of the laity, or that congeric force external to the professional world, in matters affecting the promotion and advancement of the Homœopathic science of medicine.

The Legislature of Michigan, in 1855, amended a law, to read in these words, " The regents shall have power to enact ordinances, by-laws, and regulations for the government of the university; to elect a president, to fix, increase, and reduce the regular number of professors and tutors, and to appoint the same, and to determine the amount of their salaries: *Provided*, That there shall always be at least one professor of Homœopathy in the department of medicine." Under the authority of the above law, created by the State Legislature, the regents were empowered to disburse the salaries and funds, with discretionary powers accorded to them, *provided* they, under this law, furnish at least one Homœopathic professor. It seems that there were influences which were sufficient to induce the disregard, and resist this last proviso of this law, for, though the friends of Homœopathy took such legal measures as the applying to the Supreme Court for a mandamus to compel these regents to carry out the above law, they successfully delayed such legal measures until, in 1867, when the university, needing further pecuniary aid than its income from the townships ceded to it in 1826 brought it, applied to the Legislature for this aid.

This the Legislature agreed to grant and to procure by levying a tax of one-twentieth of a mill upon the dollar of all taxable property, providing "that the regents of the University shall carry into effect the law which provides that there shall always be at least one professor of Homœopathy in the department, and appoint such professor, at the same salary as the other professors in this department; and the State treasurer shall not pay to the treasurer of the Board of Regents any part or all of the above tax until the Regents shall have carried into effect this proviso." With a view of this grant of the Legislature, subject to the order of the treasurer of their board, they established a Michigan school of Homœopathy, at a distance from Ann Arbor, the location of the University, and appointed Dr. Hempel a professor, at this school, at a salary of one thousand dollars per annum.

Evidently the intent of all State enactments was to accord to its citizens seeking an education in medicine equal opportunities within its State University, whether they be preparing themselves as Homœopathic or Allopathic practitioners, for the creating chairs of Homœopatic Therapeutics and Materia Medica would enable such students to pursue these branches, and at the same time avail themselves of the courses in anatomy, physiology, chemistry, etc., upon which both schools of medicine place due stress, and of course agree concerning in every particular; this these students could do, and also avail themselves of those valuable opportunities which every large university extends. However, by the action of the Board of Regents, this struggle for the accordance to all of equitable treatment was prolonged.

In 1875 the Legislature approved an act, which affirmed that "The Board of Regents of the University of Michigan are hereby authorized to establish a Homœopathic medical college as a branch or department of said University, which shall be located at the city of Ann Arbor. The treasurer of the State of Michigan shall, on the first day of January, 1876, pay out of the general fund, to the order of the Board of Regents, the sum of six thousand dollars, and the same amount on the first day of January of each year thereafter, which moneys shall be used by said Regents exclusively for the benefit of said department."

At last, after many years, could the University of Michigan offer to those of her citizens "who desire to be thoroughly indoctrinated into all that pertains to a scientific and systematic course of study," a course of instruction within her walls where the "requirements of a thorough scholarship" and the demands of the "Homœopathic Intercollegiate Congress" could be extended.

Doubtless these matters have been wearying to the reader, but their careful perusal will be profitable, for we have now carried him to a point from which he can judge intelligently of the action of the American Medical Association at its last regular meeting in 1881. This Supreme Council of Allopathy, which has been before now the subject of considerable attention upon our part, having in view the fact that there was a flourishing department of Homœopathy attached to the medical department of the University of Michigan, *Resolved,* "It is not in accord with the interest of the public or the honor of the profession that any physician or medical teacher should examine or sign diplomas or certificates of proficiency for, or otherwise be specially concerned with, the graduation of persons whom they have good reason to believe intend to support and practice any exclusive and irregular system of medicine."

If our readers keep before themselves the disciplinary methods of this American Medical Association, as exemplified in its treatment of the delegates of the Massachusetts Medical Society, and already set forth in these pages, it will occur to them what is the present position of the Professors of Anatomy, Physiology, and other such chairs of the Medical and Surgical Department of the University of Michigan. These professors educate in their respective branches the students pursuing also the courses which treat of Homœopathic medicine, and are therefore required by the University to examine them and certify as to their proficiency.

Now the members of the Allopathic profession, by their acts as professors in the University, in the eyes of the American Medical Association, attach both to themselves and the State and other societies of which they are members, stigma and dishonor; and we may believe are liable to have such castigation and penalties meted out to them by this Supreme Council of Allopathy as will insure a careful observance of the mandates of this council.

This reference to the course of bearing towards the new school of medicine by the American Medical Association, in the year of our Lord 1881,

18

has been made at this time that the reader may
judge if there be any change of heart in the Allopathic faculty as regards the Homœopathic profession.

That the dominant school, at the present time,
still is wont to summons to its aid, as of yore, its
ally *mockery*, it would be superfluous to take space
and effort to corroborate for the reader's benefit,
when all are in possession of competent evidence to
clearly establish this fact. Surely, if the current
literature of the dominant school can exhibit a
single periodical in which, during any year, there
has failed to appear slanders and mockeries concerning the new school in medicine, we shall will-
ingly qualify this last statement. From the time of
that prominent Allopathic authority, James T.
Simpson, M. D., down to the present, what Allopathic physician has not entertained his listeners
with some modification of Dr. Simpson's most delectable mockery upon Homœopathy? What lay-
man in medicine has not been told by ye Allopath
that Johnny Smith had reveled in a box of Homœopathic medicines, mixed them up, and finally
fired the whole down his precious neck to get rid of
them, and lo ye regular declares the upshot was

that no purging, emesis, or any drug conditions supervened, and, therefore, what a humbug and what a fraud Homœopathy must be! Or, mayhap he chose to polish off this " humbug " by quoting that delectable bit of poetry (?) which Dr. Barr Meadows gives his readers in his concluding words attacking Homœopathy:

> " Thus have we robbed *Similia* of its trappings,
> Its base assumptions and presumptuous ravings;
> And viewing thus its native nothingness—
> Behold this 'Great Something,' Naked Lies."*

We may now appreciate that the lesson which the history of medicine with the antique races teaches to future generations was completely lost upon the sectators of Dogmatism, for if the complete subversion of the offices of medicine to the interests of the priest-class was the distinguishing feature of the history of the medicine of antiquity, so it was also the reason why such marked poverty of improvement in the medical art was at this age manifest.

We say that these lessons of antiquity were without profit to Dogmatism, for if no distinctive prin-

---

* Errors of Homœopathy. Dr. Barr Meadows, Licentiate of the Royal College of Physicians, Edinburgh, etc.

ciples are found associated with the medicine of antiquity, and if the weal of the human race as regards the possibilities of the medical art were made secondary by the governing priestly class to their own interests and investitures; if the aim and effort of this autocratic class *was not* to open up the lights of science and follow the truth fearless of where it might lead them; but, on the contrary, to set these matters one side and to struggle for the possession of the power and fat emoluments that could be acquired by careful manipulation of the offices of medicine; if these are true, as we have seen, concerning the priestly orders of the antique races, they certainly are, in every sense, true of the history of the Dogmatic School of Medicine down to this day.

Abundant evidence of this has already been set forth in these pages. We have seen that it matters not, that at different times more than two hundred theories have, in their turn, been summoned by Dogmatism to represent what should be the accepted explication upon medical questions, and that after serving as the basis for the application of most injurious methods of treatment have been displaced by this autocracy for fresher and more fanciful hypotheses.

Little care the Allopathic faculty that Sir William Hamilton, one of the most prominent literateurs of his day, and so looked upon and recognized by the prominent scholars and thinkers of Great Britain at that time—little care they that this distinguished authority tells us that " Homœopathy and the water-cure are now and here (Edinburgh) *blindly* anathematized as heretical; in the next generation it is not improbable that these same doctrines may be no less blindly preached as exclusively orthodox. Such is poor human nature! Such is corporate—such is medical authority." *

The true reason, then, why the dominant school complacently views its past, and subsequently upon such experiences as are peculiarly its own suffers no apparent discomfiture, but with brazen face and positive tone reasserts that in the Dogmatic faculty power is delegated to pronounce as to what is consonant with the truth in medicine, and what is not; the disfigurement or demolition of these then matter nothing to this school, so that its chief aim, the places of power, and the power of authority, are permitted to dwell in it.

Hence, this systematic and persistent persecution

---

* Discussions, etc., page 638.

of the new school. But how short-sighted this policy of theirs is, the reader will now appreciate, for it is easy to see that the vitality of the new school in medicine rests not in the Homœopathic profession, but in a more potent element, the *laity*. How puerile such attacks upon the Homœopathic profession, when if it were possible for Allopathy to wipe out of existence, at one fell sweep, the whole Homœopathic profession, we are justified in asserting that such is the recuperative capacity the Homœopathic laity has so often demonstrated in rallying to the support of this cause as against the onslaughts of Dogmatism, as would be competent to raise up a profession to represent and employ the medical truths which they believe in, and look to when their application is required or needful.

Nor is this more than the new school in medicine deserves, for if Allopathy is the strife for power, Homœopathy has been a struggle for liberty. Thus, it will be seen that Homœopathy has wisely declined to enter the lists with Allopathy upon its issue, and is not and has not been in the effort to dispossess the dominant school of what Allopathy fain would continue to hug to its bosom; on the contrary it is a struggle with Homœopathy to obtain the right to *dissent* from the false tenets of Allopathy, and still be recognized in the exercise of this right to that extent, that humanity shall not, in any instance, suffer through the barriers which Allopathy would endeavor to place in the path of duty which the vagaries of disease make it necessary that every physician shall conscientiously follow.

We have already seen that the adherents of the dominant school in medicine exalt the doctrines of Allopathy into a medical faith, that they by directing their medical colleges to compel students, who in regular course graduate from their institutions, to declare that they will renounce the truths of the new school even before they have been investigated by such students, make their diplomas not certificates of knowledge, but confessions of faith, thus placing such colleges in a position consistent with the medical system which they represent.

This medical faith which is grounded in the ever shifting sands of popular prejudice, this Ishmael of medicine for a reason as competent as that relating to the younger half-brother in medicine will not " cross the Stygian ferry;" on the contrary, "I" * * " will make fruitful, and will multiply him exceedingly." However, at this epoch, he has been and still poses before us as a *mocker*, therefore, from the pale of civilization will he be *cast out* to dwell and exercise his peculiar offices with the paganic and barbarous races ; for the hand of every educated mind will be against this Ishmael, while this outcast will, as his instincts and predilections have ever influenced him to act in the past, continue to raise his hand against everything that opposes his assumption of power and of rule.

The civilized world will thereupon assert its preference for employing the good offices of the Isaac in medicine in overcoming disease, that man of gentle nature, a clear definition of whose exponent was so many years ago concisely set forth by one of the most prominent and eloquent authors— William H. Holcombe, M. D.—in the ranks of the Homœopathic profession.

Dr. Holcombe tells us that a Homœopathic physician is one who uses all the surgical, obstetrical, mechanical, and chemical measures where indicated, and who in the vital or dynamic sphere is guided by the Homœopathic law—   .

Similia Similibus Curantur.